CONSTRUCTING & MAINTAINING YOUR

WELL &
SEPTIC SYSTEM

No. 1654
$19.95

CONSTRUCTING & MAINTAINING YOUR
WELL &
SEPTIC SYSTEM

BY MAX & CHARLOTTE ALTH

TAB BOOKS Inc.
Blue Ridge Summit, PA 17214

FIRST EDITION

FOURTH PRINTING

Printed in the United States of America

Reproduction or publication of the content in any manner, without express
permission of the publisher, is prohibited. No liability is assumed with respect to
the use of the information herein.

Copyright © 1984 by TAB BOOKS Inc.

Library of Congress Cataloging in Publication Data

Alth, Max, 1917-
Constructing and maintaining your well and septic
system.

Includes index.
1. Wells—Design and construction.   2. Wells—Main-
tenance and repair.   3. Septic tanks—Design and con-
struction.   4. Septic tanks—Maintenance and repair.
I. Alth, Charlotte.   II. Title.
TD405.A535   1984   628.1'14   83-24124
ISBN 0-8306-0654-8
ISBN 0-8306-1654-3 (pbk.)

Also by the Author from TAB BOOKS Inc.

No. 1554   *Be Your Own Contractor: The Affordable Way to Home Ownership*

Contents

Dedicated to
Erin
Shannon
Kim
Simon
Michele
Darcy
Michael
Charlotte, and all other lovers of pure drinking water

Introduction

YOU WOULD LIKE TO OWN YOUR HOME. A year-round home or possibly just a home for the summer. Not too many years ago this was a dream many of us could realize in a dozen years or so. We could work hard, scrimp and save, and accumulate sufficient money for a down payment. Eventually, we would own our own home free and clear.

Today the difference between most peoples' income and the sum needed to purchase a home is several times greater than ever before. One way to reduce the sum needed to purchase a home and the attendant waiting period is to purchase property beyond the reach of municipal water and sewage lines. In general, rural property is less expensive than city property, and the taxes are generally lower.

Most of us have been born to homes with water at the end of a faucet and a clean toilet bowl with the turn of a handle. The thought of depending on the vagaries of one's own water and sewage system can be disquieting.

But there is no need for a feeling of uneasiness. Millions of American homes have their own water and sewage systems. Millions more are installing their own systems to reduce the initial cost of their homes. The systems work and are dependable.

This book explains and describes the workings of these systems, how you can do much, if not all, of the work yourself, and how you can maintain these systems with a minimum of time and effort. Begin by reading the entire book. Before you start construction, carefully check the percolation test as described in Chapter 20.

Part 1
Wells

Chapter 1

Groundwater

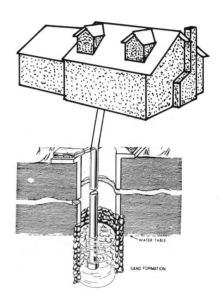

IN THE MOVIE *GIANT*, JAMES DEAN, JEALOUS OF A neighbor's wealth and beautiful wife, sets up a spring-pole well digger. By jumping up and down on the pole, succeeds in drilling down to oil and a fortune. It happened in the movies, but it has happened in real life too. Today, seekers of oil might drill 5 miles or more into the earth before perhaps finding oil.

When I was a boy I spent time on a farm in the Catskills. The farmer got his drinking water from a spring that gushed out from a crack in the hill. A hundred yards away, there was a creek with good, cold drinking water. Outside the city of Denver, neighbors chip in to hire a professional driller to go down several hundred feet to find water that they will share.

The point of all this is simply that in some places water is easy to find; it sometimes flows out of the earth in the form of a spring or an artesian well. In other areas, a lot of hard work and considerable amounts of time and equipment are required. If water is visible, there might be no problem. If there is no water in sight, how can you tell whether or not you can reach water by putting down a well; how can you tell how deeply you will have to go; how much water you can expect for your efforts?

Unfortunately none of these questions can be answered with a formula or a set of instructions or measurements. To answer these questions with any useful degree of accuracy, you must understand the nature of ground water and aquifers and study the problems the way a hydrologist does.

WATER IN THE EARTH

When you dig down into the earth, you pass through a layer of earth hydrologists call the zone of aeration. This zone or layer consists primarily of soil. The top of this zone is the surface of the earth, and the bottom of the zone can be described easily, but not accurately as resting on the water table.

There is almost always a little water to be found in the zone of aeration. Most of this water is found in smaller openings. The larger openings generally contain air. Following a heavy rainfall this zone will become saturated with water. See Fig. 1-1.

3

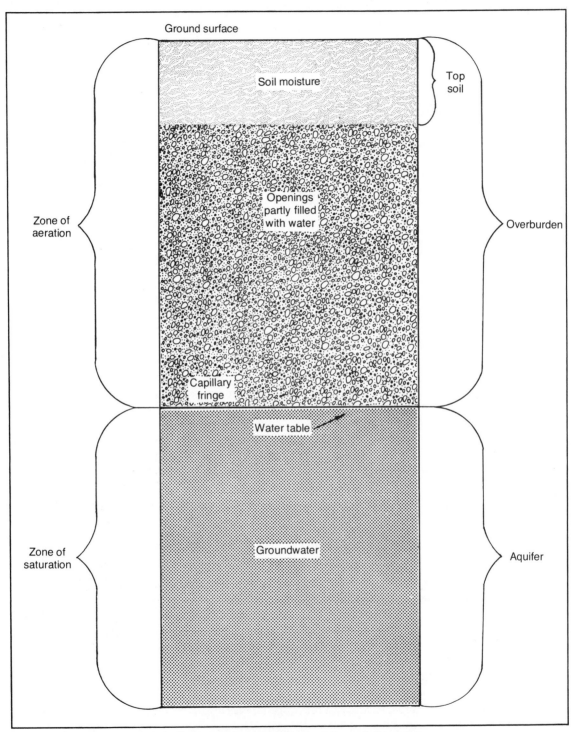

Fig. 1-1. A hydrologist's view of the first several hundred feet of earth beneath the surface.

If there is an extended dry spell, most of the water will be removed by surface evaporation and transpiration (plants drawing water from their roots up into their leaves). The water in the zone of aeration (Fig. 1-2) will not enter a well. In other words, digging a hole into moist or damp earth will not result in a flow of water into the hole. Within the zone of aeration, water is held in place by the attraction between molecules of water and adjacent particles of soil and sand called *capillarity*. This is why you can turn a glass over and shake it hard and yet not get it perfectly dry; a film of water will always adhere to the sides of the glass.

Groundwater. Groundwater is found beneath the zone of aeration. The uppermost level of the groundwater is called the *water table*. Groundwater sometimes takes the form of an underground stream flowing slowly in a dark tunnel. It might even be in the form of a lake such as in Howe's Cavern in upstate New York or the limestone natural caves in Kentucky. But they are rarities. Almost all the groundwater we know of is found in tiny cracks and pores in the rocks beneath the aeration layer. Groundwater is found in fractured rock, sand, sandstone, and gravel.

Rains fall upon the earth where some of it is absorbed by the zone of aeration. When the thirsty plants and the soft earth have absorbed all the water

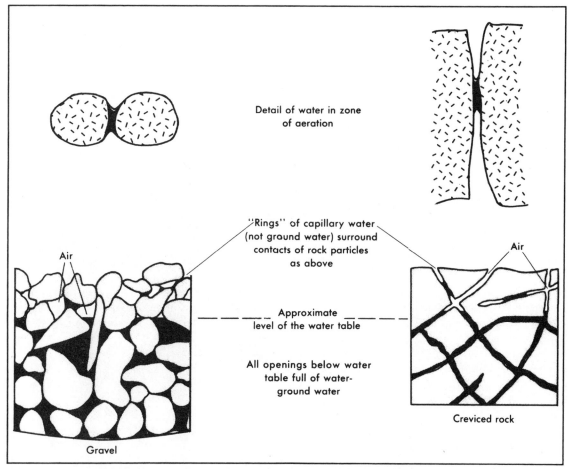

Detail of water in zone of aeration

"Rings" of capillary water (not ground water) surround contacts of rock particles as above

Air

Approximate level of the water table

All openings below water table full of water-ground water

Air

Gravel

Creviced rock

Fig. 1-2. Top: Particles of earth in the zone of aeration are so close to one another that capillary attraction holds the water in place. Below: The spaces between the particles of stone are sufficiently apart in the zone of aeration to permit the water free movement.

they can hold, the remaining water continues to either run off or move down into the earth where it joins the rest of the groundwater. The addition of more water to the existing groundwater raises the level of the water table. That is to say, the level of the water in the pores, cracks, and crannies in the rock or sand is raised.

The cracks extend so far down and no further. The fractured rock, sand, or gravel zones contain *aquifers*. *Aquifer* is derived from two Latin words: *aqua*, meaning water, and *ferre*, to bring. An aquifer brings underground water.

The reason a hole dug into an aquifer will turn into a well (collect water) is that the openings between the pieces of sand or gravel or shattered stone are so large that capillary attraction is negligible. See Fig. 1-3.

The thickness—the distance from the top to the bottom of the aquifer—varies with the nature of the earth. The aquifer might only be a few feet thick or it might be hundreds of feet thick. It might extend for miles without a break or it might stop abruptly because of some prehistoric geological upheaval. Generally, the unseen aquifer follows the topography of the land's surface. Where the land dips the aquifer dips.

The flow of water generally follows the topography of the land. Groundwater will flow downhill in the same direction the surface of the land descends. The movement of groundwater through the aquifer is slow as it curves around barriers posed by solid rock. The water will take easier routes through gravel and coarse sand. Eventually the groundwater might emerge as much as 50 miles away as a spring or join a stream, unnoticed. See Fig. 1-4.

Water can travel through gravel at a rate of hundreds of feet a day. In fine sand or silt, water might not move more than a few inches a day. While the flow of an open stream can be measured in feet per second, the flow of groundwater is usually measured in feet per year. Still, in the northern Great Plains, tests have shown that underground water travels hundreds of miles through the Dakota Sandstone.

There is no firm relationship between aquifers and the depth at which they exist. In New England,

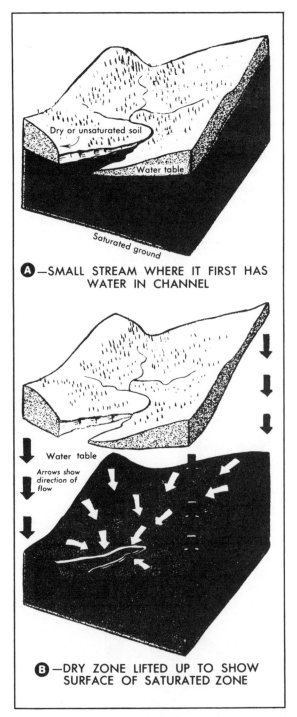

A—SMALL STREAM WHERE IT FIRST HAS WATER IN CHANNEL

B—DRY ZONE LIFTED UP TO SHOW SURFACE OF SATURATED ZONE

Fig. 1-3. Underground water in the saturated zone (aquifer) flows in a generally downhill direction.

6

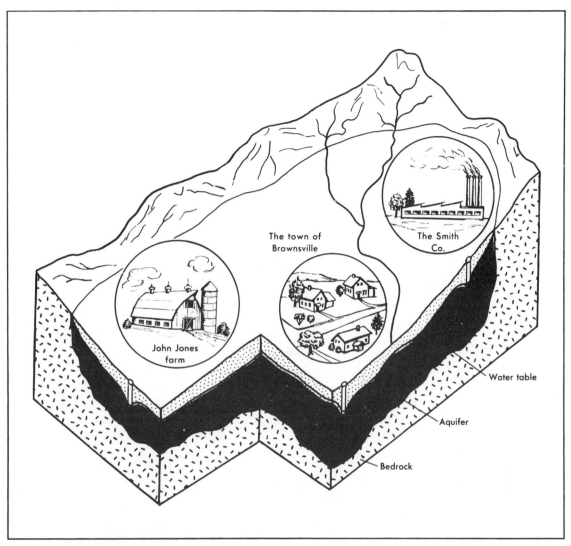

Fig. 1-4. A single aquifer can supply a very wide area with water.

for example, there is very dense granite to be found at or near the earth's surface. In the Great Plains, there is very porous sandstone lying several thousand feet beneath the surface. Generally, however, the further down you go the less porous and less permeable are the rocks you find. Fresh water has been found at a depth of 6000 feet and salt water has been found along with oil at depths of more than 20,000 feet. As a rule, few wells drilled deeper than 2000 feet find water. The weight of the rock closes all the cracks, crevices, and crannies.

RECHARGING

Groundwater is normally recharged or replaced by rainfall. The percentage of the rain that replenishes the supply of groundwater depends upon the nature of the soil upon which the rain falls and its grade (pitch). If the soil is porous and relatively flat, most of the rain will seep down into the aquifer. If the soil is hard and has a high percentage of clay and, in addition, if the land's surface is steeply pitched, most of the rain will run off and not reach the aquifer.

For a well digger, this means that drilling a well at the bottom of a large hill does not guarantee the finding of a dependable quantity of water. There might be an aquifer within relatively easy drilling distance, but if the surrounding area does not permit recharging there will be little dependable water in the well. Instead of recharging the aquifer, the rain will come down the hill in a troublesome surface runoff stream. If the soil on the hill is porous but overlays a solid layer of clay, there will be less surface runoff but the recharge rate will not be much better. See Fig. 1-5.

WATER QUANTITY

The quantity of water that can be more or less continuously drawn from a well depends upon the permeability of the aquifer, its size, the recharge area, rainfall in the recharge area and the length of

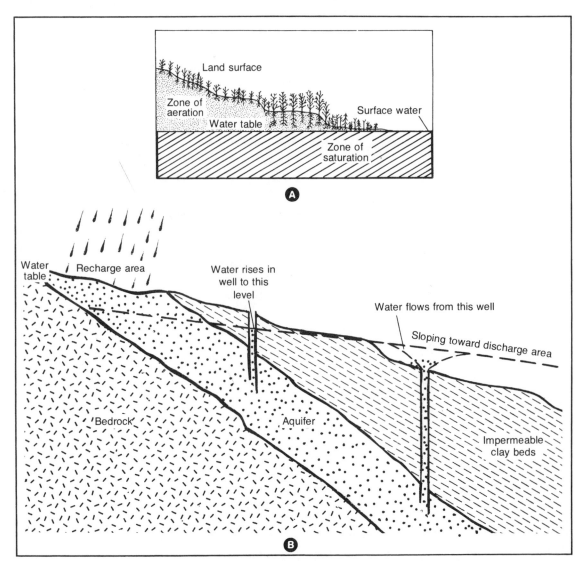

Fig. 1-5. A: The relation of surface water to the water table. B: Rainfall can recharge an aquifer. Note how the water table appears in the normal well and in an artisian well (well at right).

the well screen in the aquifer. To a very small extent, the diameter of the well also affects the quantity of water that can be drawn. The larger the diameter the more cracks the well intersects and the greater the quantity of water that will enter the well. Should the water table level drop below the bottom of the intake end of the well, no water will be available from that well no matter what its diameter.

Aquifers. The thickness of an aquifer might be no more than a few feet or it could be several thousand feet. The aquifer might extend no more than a few hundred feet in any direction or it might cover an area of several hundred square miles. The Dakota Sandstone carries water across several states. Nevertheless, a great many aquifers are under a mile or two across.

All aquifers rest on solid rock that is watertight. The rock might be found at depths of a few hundred feet or as much as several thousand feet below the surface. The quantity of water that can be contained in a number of cubic feet of aquifer depends upon the porosity of the aquifer. *Porosity* is the spaces, crevices and cracks between the pieces of rock that can be filled with water that can't be held in place by capillary attraction.

In the case of sand, which is considered unconsolidated rock, the porosity—the open spaces between the grains—can amount to 30 or 40 percent of the total volume. These spaces can easily contain water that can move as dictated by gravity. The porosity of sand and gravel depends not only on the size of the pieces of stones but on their relative sizes. When the grains are all about the same size, well sorted as geologists say, the grains do not pack as tightly as when they are poorly sorted. In the latter case, the smaller grains tend to slip into the spaces between the larger grains, or stones in the case of gravel, and the result is less air space and less porosity.

When the aquifer is highly permeable and is exposed to recharging rain, the aquifer can supply large quantities of water to the well. If the aquifer is not very porous, the cracks are relatively few and not well connected, water cannot flow easily and wells in this type of aquifer will perform poorly.

Different types of rocks vary widely in porosity. In some very dense rocks such as granite, porosity will be less than 1 percent. The other end of the scale reaches 40 percent for sand, gravel, and unconsolidated rock. In some instances, the absence of an appreciable quantity of pores does not stop the flow of underground water. Some compact, consolidated rocks such as granite and slate will have sufficient cracks, called *joints* by geologists, to permit the flow of groundwater. When and where these joints intersect, water can flow almost as easily as through an ordinary water pipe. Wells are practical in such consolidated, dense stones only when there are sufficient cracks and the well intersects with a goodly number of them.

Groundwater moves much more rapidly through some kinds of rocks than others. Groundwater moves readily through sandstone because this type of stone has natural pore spaces. The flow of water through solid granite, schist, or slate is negligible—there are no cracks. Clay and silt are almost permanent barriers to water movement. On the other hand, coarse gravel provides free and easy passage. Limestone is often cavernous; water very often flows through limestone much faster than it does through other formations.

There appears to be no fixed relationship between the porosity of rocks and their depth beneath the surface of the earth. In general, if you go down far enough you will always reach solid, watertight rock formations. Nevertheless, you can find granite coming up through the earth's surface in New England and ajoining states, and water-bearing sandstone several thousand feet below the surface of the Great Plains.

GROUNDWATER RESERVOIRS

Some geologists use the words groundwater reservoir and aquifer interchangeably, but generally the term *groundwater reservoir* describes the zone of saturation. Assume that a particular aquifer covers an area of 10 acres. Assume that it is possible to measure the total height of the water in the aquifer from the surface of the solid rock to the water table. This body of water, held in suspension in the cracks and pores of the aquifer, is the groundwater

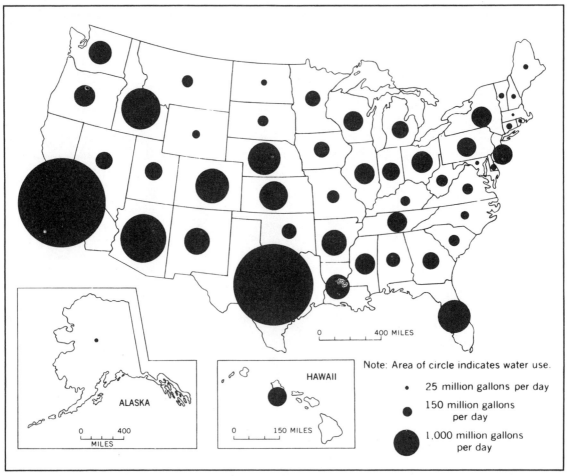

Fig. 1-6. Total groundwater use in the United States during 1960.

reservoir—the zone of saturation. See Fig. 1-6.

It is from all the groundwater reservoirs combined that we draw our well water. The total quantity runs into the billions of gallons a day. No one knows exactly how much water there is in our groundwater reservoirs, but it is estimated that our underground water supply is several times greater than all our fresh-water lakes and surface reservoirs combined. This sounds like a great deal of water, but we are running short and our government is taking steps to conserve and replenish groundwater.

One reason is usage. We are simply removing water faster than nature is replenishing it. Another is the nature of a typical groundwater reservoir.

When you have water in a lake or an artificial storage pond, all the water in the pond is immediately available. You can remove the water as fast as you can pump it out. Underground water is contained in billions of pores and cracks. The rate of water flow from the far reaches of the aquifer to the well and pump is limited. The rate at which you can withdraw water from an aquifer is therefore limited. Thus when you pump the water out of a well more rapidly than the groundwater can enter the same well, your well will run dry (although there might be lots more water in the aquifer).

Cones of Depression. Because water does not flow freely through an aquifer, when water is drawn from an aquifer by means of a well the water

10

table (water level) falls in the area immediately surrounding the well. Draw water from a lake or pond and the entire surface drops. Not so in an aquifer. The area in which the water table drops around a well in use is called a *cone of depression*.

The shape of the cone depends on the nature of the aquifer, whether it is highly permeable or only slightly permeable, and the rate at which water is drawn from that particular well. Draw water too rapidly from a well and the well might go dry or show reduced flow even though there is still plenty of water nearby. When two or more wells draw from the same aquifer, the maximum rate of flow into each will be reduced. The cone of depression formed around one well will extend to the adjoining well and so reduce the quantity of water there.

SPRINGS AND ARTESIAN WELLS

Springs are formed when the lower end of an aquifer extends through the surface of the earth. Very often the aquifer extends through the side of a hill. From there, the aquifer is beneath the zone of aeration angling up toward the top of the hill. Rainwater penetrates the zone of aeration and seeps downward, escaping through the exposed end of the aquifer. Sometimes the aquifer is an unconsolidated

Table 1-1. Practical Depths, Usual Diameters and Geological Formations Suitable to Different Types of Wells.

Type of well	Depth	Diameter	Geologic formation
Dug	0 to 50 feet	3 to 20 feet	*Suitable:* Clay, silt, sand, gravel, cemented gravel, boulders, soft sandstone, and soft, fractured limestone. *Unsuitable:* Dense igneous rock.
Bored	0 to 100 feet.	2 to 30 inches.	*Suitable:* Clay, silt, sand, gravel, boulders less than well diameter, soft sandstone, and soft, fractured limestone. *Unsuitable:* Dense igneous rock.
Driven	0 to 50 feet	1¼ to 2 inches.	*Suitable:* Clay, silt, sand, fine gravel, and sandstone in thin layers. *Unsuitable:* Cemented gravel, boulders, limestone, and dense igneous rock.
Drilled: Cable tool	0 to 1,000 feet.	4 to 18 inches.	*Suitable:* Clay, silt, sand, gravel, cemented gravel, boulders (in firm bedding), sandstone, limestone, and dense igneous rock.
Rotary	0 to 1,000 feet.	4 to 24 inches.	*Suitable:* Clay, silt, sand, gravel, cemented gravel, boulders (difficult), sandstone, limestone, and dense igneous rock.
Jetted	0 to 100 feet.	4 to 12 inches.	*Suitable:* Clay, silt, sand, ¼-inch pea gravel. *Unsuitable:* Cemented gravel, boulders, sandstone, limestone and dense igneous rock.

rock and the water seeps from exposed cracks. Depending upon the size and nature of the aquifer, the spring might be an all-year dependable source of good water or it might run only following a rain.

Artesian wells are so-called after the ancient town of Artois in France, which was formally the old Roman city of Artesium. During the Middle Ages, it was there that the best-flowing, or at least best-known artesian wells, were to be found.

Artesian wells differ from springs in the pressure that is behind the water that issues from the ground. With a spring, the water merely flows out of the aquifer. To be classified as artesian, the water must emerge under pressure. The pressure might be low, in which case it merely bubbles out, or it might actually gush and shoot up into the air.

The mechanism is easy to understand. The flow of water from an artesian well is similar to the squirt of water that emerges from a hole in a water hose filled with water under pressure. In the case of the artesian well, the aquifer reaches high up a hill or mountainside. Where the well is drilled, the aquifer is overlain by a layer of impermeable rock or clay. The water in the aquifer is under pressure. When a hole is drilled through the impermeable layer, water pressure shoots the water out of the well and into the air. Deep wells are sometimes called artesian, but unless the water in these wells comes up of its own accord they are not artesian. See Table 1-1.

Chapter 2

Water
Quantity and Quality

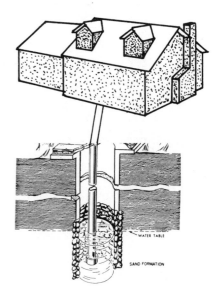

THERE IS NO CERTAIN WAY OF FORECASTING THE quantity of water that can be drawn from a well before it is dug or drilled. If you calculate your water needs and learn what quantity of water can be drawn from neighboring wells, you can estimate your chances of securing the quantity of water you need. If your neighbor's flow rate is far less than what you will require, going down farther might help. Then again you might have to install a second well. In any case, it is important that you know your needs and the chances of meeting them before you drill.

RATE OF CONSUMPTION

The common figure for water consumptions is usually given at 50 to 60 gallons per day per individual. This might have been true some years ago, but today with dishwashers and clothes washers found in almost every home, these figures are much too low. See Table 2-1.

Additional Human Consumption. The estimated figure of 95 gallons of water per day per individual in a modern household does not include water used to wash the car or fill the pool. That would be additional water. And, of course the figure would be multiplied by the number of adults usually present in the house. When and where there would be guests, the figure would have to be increased accordingly.

Lawn and Garden Water Consumption. The usual figure published for water requirements by a lawn and by a garden is 2 inches of water per week. This might appear to be one heck of a lot of water, but on an average you need 120 gallons of water to produce one dry, edible pound of plant. In other words, a pound of dry beans cost you 120 gallons of water.

Two inches of water works out to 1.5 cubic feet of water on every square yard of surface. One cubic foot of water equals 7.48 gallons of water. Thus you need 11.2 gallons of water for every square yard of your lawn or garden's surface. Don't forget rain. From local records you can learn the average rainfall in inches that you can expect. That will give you

Table 2-1. Water Consumption Rate.

Consumer	Quantity per 24 hours
Human	10 gallons
Cow	7 to 15
Horse	5 to 10
Hog	12
Sheep	1
100 chickens	4
Lavatory	1.5 Per one-time use
Bathtub	30 Per one-time use
Toilet, tank type	6 to 8 Per one-time use
Shower bath	6 to 8 Per one-time use
Dishwasher	25 to 35 Per one-time use
Clothes washer	30 to 40 Per one-time use

Thus:

1 person per day uses

10 gallons cooking, etc.
 6 gallons in the lavatory
30 gallons in the tub
24 gallons using the toilet
15 gallons using the dishwasher once every 2 days
10 gallons using the clothes washer once every 4 days
——
95 gallons per day

a ballpark figure. With this figure and the known area of your lawn and garden, you can compute your "farming" water needs. To keep accurate tabs on the rainfall, place a straight-sided, open-top can on an out-of-way corner of your property. Then measure the depth of the water collected after every rain.

Providing the Needed Quantity. Assume that there are four adults in your family and that you rarely have guests. On the basis of the figure given, you will require 4 × 95 gallons of water per day for a total of 380 gallons of water per day. This works down to an hourly rate of 15.8 or 16 gallons of water per hour that must be supplied by your well if you are not going to run out of water.

The chances that you and your family are going to use an even 16 gallons of water per hour are slim. There is always the possibility that someone will be using the tub while someone else is washing clothes, and someone will be drawing water for washing or whatever. Thus water consumption will rarely be evenly spaced through the 24 hours. There will be many times when several valves and faucets will be opened simultaneously. When this occurs, the well must supply the total immediate demand or water will flow at half volume or less into the various plumbing fixtures and appliances.

Referring to Table 2-2, you can see that the optimum required flow rate for a bathroom tub, dishwasher and clothes washer amounts to a total of

Table 2-2. Recommended Flow Rates and Minimum Pressures for Common Household Fixtures.

Type of Fixture	Minimum Required Pressure	Desired Flow Rate
Lavatory faucet	8 psi	3 gpm
⅜ sink faucet	10	4.5
½ sink faucet	5	4.5
Bathtub faucet	5	6
½ laundry tub	5	5
Dishwasher	8	5
Clothes washer	8	5
Shower	10	5
Toilet tank	10	3
Toilet flush valve	15	15-40
Sill cock plus 50′ Garden hose	30	5

16 gpm (gallons per minute). That does not leave any water (at proper pressure) for the remaining faucets and fixtures. The problem cannot be solved at this time, but an answer can be approximated.

Start by estimating the probable flow rate from the well you are still *planning* to dig or drill. Do this by learning the constant available flow rate that is produced by a nearby well. This figure assumes that your well will reach the same aquifer and that both wells will not be so close to one another that the cone of depression around one will adversely affect the other.

Assume now that half the fixtures and faucets in your home might be turned on at one time. Using the same table, you will come up with a figure of 28.5 gpm, which is just an educated guess.

If you estimate that your well will produce 30 gpm, then in theory it will never run dry. You require no more than a minimum-size water storage tank (just to be on the safe side). See Fig. 2-1.

If you estimate your projected well will produce 20 gph, there certainly will be periods when the demand exceeds the supply. As stated in this example, there will be times when you will need water at a rate of 28.5 gpm and you will need it long enough to fill the bathtub, dishwasher, clothes washer, toilet, and other fixtures. All in all, using the table you are going to require some 100 gallons plus of water in less than 5 or 6 minutes.

The number of gallons is a little high because it doesn't consider the flow of well water into the system during the 5 or 6 minutes required to fill the tub or washer. Also, washers do not take their water at one gulp, but over a time period of up to 30 minutes or so. In any case, this is the way you can estimate your water requirements, your water tank size, and just how well you can expect your water well to fulfill your needs.

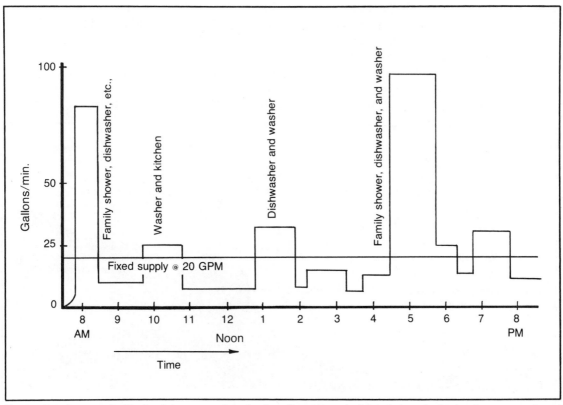

Fig. 2-1. Drawing a graph of the family's consumption of water will enable you to determine required water storage capacity and average need.

WATER PURITY

From our point of view it is important to evaluate water as having two characteristics: purity and quality. Hydrologists usually consider purity a function of water quality.

Pure Water. We define pure water as water fit and safe to drink. Water might be crystal clear, cold and even free running, and yet it may contain pathogens, microbes and bacteria injurious to one's health. Generally, if your well is 50 to 100 feet away from the nearest source of contaminations, chances are fairly good that the water will be safe to drink. The earth is a tremendous filter and purifier. In order to be safe, to be certain, always have your water supply tested.

Your Local Board of Health. Your local health or sanitation department stands ready and willing to check your water supply at little or no cost. Either an inspector will visit your home site and inspect the land for the presence of insecticides, the location of your proposed well in relation to the septic tank, barn and the like, or he will provide you with a sterile bottle and instructions on how to fill it (assuming the well is in).

They will make a bacterial and coliform count, which indicates the quantity of human and other warm-blooded animal fecus present in the water. They will then inform you if the sample passes their standards, and if it doesn't what steps you can take to correct the problem.

WATER QUALITY

Water does not remain in one place very long. It is in constant circulation from the land to the sea and back to the land again. See Fig. 2-2. This constant movement is called the great water cycle or *hydrologic cycle*.

The oceans, which cover three-fourths of the earth's surface, contain an estimated 329 million cubic miles of salty water. Each day some 210 cubic miles of water enter the atmosphere by means of evaporation. Of this, about 186 cubic miles of water falls back into the seas as rain.

The winds blow 24 cubic miles of water over the land where it falls as rain. But the total quantity of rain, dew, and snow that falls on the land each day amounts to 62.5 cubic miles. The difference is made up by the 38.5 cubic miles of water that evaporates from the land.

The difference between what falls on the land and what is evaporated is made up of rivers and streams that empty into the oceans. This amounts

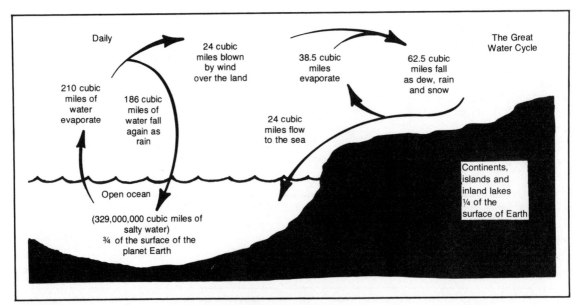

Fig. 2-2. The cyclic movement of water from sea to land and back to sea.

Table 2-3. Comparative Concentrations of Salt in Water From Different Sources.

Source of water	Salt concentration (ppm)
Distilled	0
Rain	10
Lake Tahoe	70
Suwannee River	150
Lake Michigan	170
Missouri River	360
Pecos River	2,600
Ocean	35,000
Brine well	125,000
Dead Sea	250,000

to 24 cubic miles of water on an average each and every day of the year.

Water has the greatest surface tension of all common liquids, with the exception of mercury. Surface tension is the nature of a liquid to try to pull itself into a ball (a drop of rain) and to adhere to a surface. When water strikes a leaf it spreads out. When water falls upon the earth, capillary action pulls the water into the soil and spreads it out horizontally. If it wasn't for this characteristic of water, a far greater percentage would run off the surface and directly back to the seas instead of finding its way into aquifers and nourishing the roots of plants.

The strangest characteristic of water is its ability to dissolve a tremendous number of different substances. No other known liquid comes as close as water to being a universal solvent. What is most important to us is that water itself is not changed by the substances it dissolves. By one method or another, the water can always be recovered and used again. This is a remarkable property because water can dissolve so many substances. For example, most of all the known elements have been found in sea water some as mere traces and others as relatively large quantities. For example, rain has about 10 ppm (parts per million) salt content. A lake might have 70 ppm. Some rivers can have 360 ppm. The ocean has 35,000 ppm and the Dead Sea contains 250,000 ppm of salt. See Table 2-3.

Differences in Water. We all know that surface water, the water found in rivers, lakes and oceans, differs considerably from one source to the other. See Fig. 2-3. The water of Soap Lake in central Washington feels soapy. Milk River in Montana looks milky. Stinking Water Creek in Nebraska does have a smell. Some lakes are sweet and some lakes are salty. If you travel, you know that tap water in different cities tastes differently. This is due to the difference in the quantity of dissolved minerals in the water and their nature.

Pure water, distilled water, with no dissolved minerals not only lacks taste it tastes poorly. The taste of the water you will draw from your well will depend upon the nature of the aquifer and the distance the water travels through the aquifer. When the well taps into several interbedded aquifers, the quality of the water will vary from time to time, depending upon the particular aquifer most of the water may be flowing through at the time. In some areas, the depth to which the well descends also affects the quality of the water. In northwestern North Dakota where the earth contains sand, shale, siltstone, clay, and lignite, a relatively shallow well will produce hard water. But in the same location, a well that reaches much farther down into the earth will produce soft water.

Hard Water Versus Soft Water. Hard water is a general term that denotes water that does not readily form lather with ordinary soap. Any water that has a relatively high percentage of dissolved minerals is classified as hard. Sea water, for example, is hard water.

In addition to making washing difficult, hard water poses many additional problems. Hard water tends to deposit minerals in pots and pipes. The rate of deposition depends upon the percentage of minerals in the water and the temperature of the surface that the water contacts. Thus you will find hard water forming a layer of calcium on the inside of the tea kettle and the inside of the water boiler. Little is deposited within the cold water pipes.

The tea kettle presents little problem. Usually, it is impractical to chip the stone like layer free. The kettle is discarded and replaced. The hot-water pipes present a different situation. Even a thin layer of stone reduces heating efficiency. A thick layer, and it can grow to ½ inch or more,

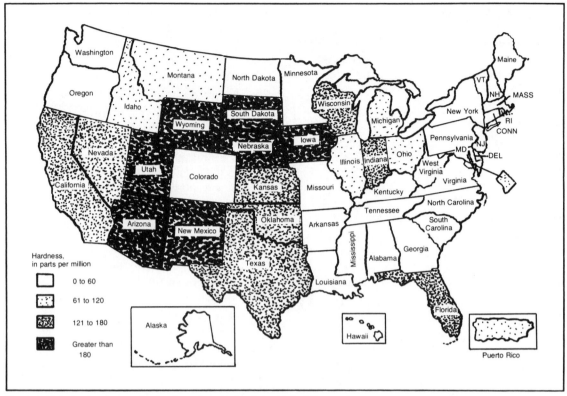

Fig. 2-3. Relative hardness of water in various parts of the United States.

practically insulates the water in the boiler from the applied heat. The only practical solution is the replacement of the boiler and associated pipes. In some instances, a hydrocholoric acid treatment helps a little. The wholesale replacement of the boiler and associate pipes can be costly. One recourse is the installation of a water softener. This is a tank and pipe system that requires little attention and maintenance and removes most if not all the minerals from the water.

In addition to the cost of the water softener and the annual or semiannual service they require, the water softener introduces a small quantity of salt (ordinary table salt) into the water as well as removing the minerals. Additional piping can be added to supply the kitchen with water that has not been softened. All this requires is a pipe fitting and a length of pipe that circumvents the softener.

To determine the hardness (Table 2-4) or softness of your water supply—in addition to trying to

estimate it by making a lather with it—you can check with your neighbors, ask the local water-softening company (they usually provide free testing) and you can bring a sample to your local board of health. Generally, they do not charge or charge only a small fee.

Dissolved Minerals. The concentration of dissolved minerals in surface water tends to vary with time and the seasons. For example, the saline content of the Great Salt Lake in Utah was about 15 percent during the 1870s, when the water in the lake was high, to 28 percent in the early 1900s and

Table 2-4. Hardness Comparison.

Parts per million	
0-60	Soft
61-120	Moderately hard
121-180	Hard
More than 180	Very hard

18

1960s when the level of the water in the lake was low.

Groundwater mineral concentration will vary with time, but only slightly. The change is never nearly as great as that found in surface bodies of water. The quantity of dissolved minerals and their types are very important. While the major causes of hardness in water are calcium and magnesium, other minerals such as iron, aluminum, manganese, barium, strontium, and free acid also contribute to a water's hardness. In general terms, hardness is usually measured by the quantity of calcium carbonate (limestone) or its equivalent that is left behind when a specific amount of water is evaporated.

Water with less than 60 ppm of dissolved minerals is rated as soft and is suitable for all purposes without treatment. Water in the 61 to 120 ppm range is termed moderately hard. Water with this range of hardness can be used for most applications with little difficulty. Exceptions are high-pressure steam boilers and other applications involving high temperatures. When the hardness range is above 121 ppm, you usually need some type of softener or special soap. Water in the hardness range of 180 ppm and over requires water-softening treatment. If the minerals are not removed, they will quickly clog the interior of the pipes and your cooking utensils. Hard water is also tough on hands and clothing (shortening fabric life).

On the other hand, hard water is desirable for irrigation because it is readily absorbed by the soil. Soft water tends to form surface puddles.

EFFECT OF DISSOLVED MINERALS

Table 2-5 provides an example of the mineral constituents of various sources of water. Listed are the types of minerals and their relative percentages in ppm.

Table 2-5. Typical Minerals Found in Water.

Constituent or property	Analyses of water, in parts per million, except for pH and color			
	River [1]	Well [2]	Canal [3]	Lake [4]
Silica (SiO_2)	5.4	41	6.6	11
Iron (Fe)	.11	.04	.11	.10
Calcium (Ca)	9.6	50	83	2.9
Magnesium (Mg)	2.4	4.8	6.7	9.5
Sodium (Na)	4.2	10	12	8,690
Potassium (K)	1.1	5.1	1.2	138
Carbonate (CO_3)	0	0	0	3,010
Bicarbonate (HCO_3)	26	172	263	3,600
Sulfate (SO_4)	12	8.0	5.4	10,500
Chloride (Cl)	5.0	5.0	20	668
Fluoride (F)	.1	.4	.2	--------
Nitrate (NO_3)	3.2	20	1.3	5.8
Dissolved solids	64	250	310	25,000
Hardness	34	145	235	46
Color	7	0	97	2
pH	6.9	7.9	7.7	9.8

[1] Stream in Connecticut. [3] Drainage from the Everglades, Fla.
[2] Logan County, Colo. [4] North-central North Dakota.

Silica, S_1O_2. You find this mineral in quartz, sand, feldspar and practically all other rocks and many other minerals. Surface water generally contains less than 5 ppm of silica, but there are a few with almost as much as 50 ppm. Groundwater usually contains more silica than surface water. Silica is objectional in water because it speeds the formation of boiler scale and forms interfering clumps on the blades of steam turbines.

Aluminum, Al. You will find aluminum in quantity only in water that has been in contact with bauxite and shales that contain high percentages of this metal. When the water contains a high organic content, aluminum will be found in concentrations of 1 ppm. Acidic water also can contain considerable aluminum in solution. Aluminum is only troublesome in the feed water of high-temperature boilers and steam turbines.

Iron, Fe. Iron is one of the three most common elements found in the top 50 miles of the earth's crust. Naturally, iron is found in many rocks and soils (these soils are red in color). Unless the surface water is acidic, it rarely contains more than a few tenths of a ppm. If the water is acidic, it will often contain large quantities of dissolved iron. Groundwater, on the other hand, might contain several ppm of iron.

When the iron in nonacid water is exposed to the air, the iron oxidizes and settles out of solution. It takes very little iron to become visible. The concentration can be as low as 0.3 ppm and the iron will form red-brown stains on white china, plumbing fixtures, and clothing washed in the iron-containing water.

Manganese, Mn. This is another troublemaker. Like iron, concentrations of manganese as low as 0.2 ppm can produce dark-brown or black stains in fabrics and laundry fixtures. Manganese resembles iron in its chemical behavior. It is dissolved in appreciable quantities from rocks in certain sections of the country. Water impounded in dams will often remove the manganese from the mud at the bottom of the reservoir. Manganese is often found in water containing relatively high concentrations of iron.

Calcium, Ca, and Magnesium, Mg. These two minerals are almost always found in water. That is because calcium and magnesium are found in almost all rocks and soils. The highest concentration, however, results when the water has been in contact with dolomite, limestone, and gypsum. These two minerals are mainly responsible for the formation of scale within boilers, hot-water heaters, and associated piping.

Sodium, Na, and Potassium, K. Water that has not been distilled contains some quantity of sodium and potassium. When the concentration of these two elements is under 50 ppm, the usefulness of the water is not affected adversely. When the concentrations goes above 100 ppm, the water will foam in a steam boiler. When the concentration is higher still with a high proportion of sodium, the water should not be used for irrigation.

Carbonate, CO_3 and Bicarbonate, HCO_3. Groundwater rarely includes carbonate and bicarbonate. When water is treated with lime, some carbonate will be released and dissolved. Bicarbonate also results when water containing dissolved carbon dioxide makes contact with rocks. If the rock is granite or of similar composition, the concentration of bicarbonate rarely exceeds 25 ppm; at most it is never higher than 50 ppm. When water containing carbon dioxide gas contacts carbonate rocks the concentration of bicarbonate in the water can rise to 500 ppm. When the concentration of carbonate and bicarbonate is very high, the water is classed as alkaline. This is usually expressed as ppm calcium carbonate.

Sulfate SO_4. Very little sulfate is generally found in rivers and wells. You find this mineral mainly where the water contacts beds of gypsum and shale. It is also found in acid-mine drainage and is formed by the oxidation of sulfides or iron. When sulfate is present in large quantities, it produces very hard scale within boilers.

Chloride, C1. Chloride can range as high as several hundred ppm in stream and lakes in semiarid regions and streams that carry irrigation runoff. Sewage also acts to increase the chloride content of water. Most rock contains a percentage of chloride that can be dissolved by water. But chloride concentrations of less than 25 ppm usually present no problems. When the concentration is much higher, and the water also contains calcium and magnesium, the corrosive power of that water is greatly increased.

Fluoride, F. The concentration of fluoride found in rocks is about as great or as small as that of chloride. There is very little fluoride to be found in surface water. A little fluoride is supposedly a dental caries preventive. According to the California State Water Quality Control Board, water containing less than 1.0 ppm of fluoride will seldom cause spots on the teeth of children. Adult teeth will stay bright with flouride concentrations up to 4 ppm and this concentration is not likely to cause endemic cumulative flourosis and skeletal effects.

Nitrate NO$_3$. When the concentration of nitrate in water approaches several ppm in water, it is a strong indication the water has been contaminated by sewage or some other organic matter. Nitrate is considered the final oxidation of matter than contains nitrogen, and that includes all organic substances. Nitrate-containing water can be used for most all industrial applications, but its use is dangerous for human and animal consumption. Studies by various research scientists and medical officers have found that excess nitrate in drinking water is the main contributing factor in an infant disease best known as "blue babies." A report by the National Research Council states that water with a nitrate concentration of over 44 ppm is unsafe for infant feeding.

Trace Elements. We do not fully understand the place of trace elements in human, animal, and plant existence. See Table 2-5. We know that we need copper, cobalt, and zinc in our drinking water, and that quantities of these elements in parts per billion are necessary to health and even life. At the same time, traces of lead, arsenic, and beryllium are poisonous to man.

Turbidity. When water appears to be cloudy or a light cannot be directed through the water without considerable loss of light, the water is turbid. It is caused by fine particles in suspension. The particles might be silt, clay, sand or of organic nature. In any case the water is not clear. Generally, when the turbidity exceeds 5 percent the

Table 2-6. Trace Elements Typically Found in Public Drinking Water.

Some trace elements in tap water, Kansas City, Mo., July 26, 1961	
Element	Parts per billion
Aluminum	58
Chromium	1.2
Copper	7.4
Nickel	3.6
Lead	5.2

water is filtered before it is consumed by humans.

Taste and Odor. Bad tastes and smells in water are usually caused by decaying organic matter and often indicate water that is unsafe to drink. Sometimes, a very high concentration of minerals will also produce a bad taste or odor.

Dissolved Solids. All the minerals that can be found dissolved in water are grouped under the heading of dissolved solids. Excluding the presence of a high concentration of toxic elements, water having no more than 500 ppm of dissolved solids can usually be used for all applications. Some special industrial processes cannot tolerate dissolved solids exceeding 100 ppm. When the concentration of dissolved solids exceeds 1000 ppm, the water is termed *saline*. When the concentration is higher than 35,000 ppm, the water is classified as *brine*.

ARTIFICIAL RECHARGING

When rain falls onto the earth, seeps through the zone of aeration, and enters an aquifer, the aquifer is recharged naturally. This, of course, is the usual way water enters an aquifer and reaches a well or spring. There are, however, times when it is advantageous to aid nature in recharging. See Figs. 2-4 and 2-5.

Assume that your well is going dry, but there is a free-flowing river or creek nearby. The location and nature of the aquifer that feeds your well is insulated by clay or rock from the moving stream. You can recharge your well by drilling a second well 50 or so feet away from the first. Water from the creek is now pumped *into* the second well. Doing so recharges the first well. The reason the creek

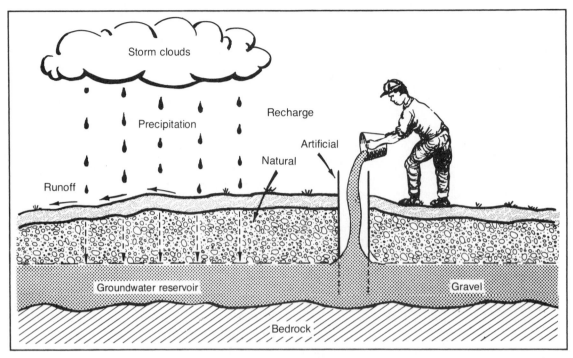

Fig. 2-4. When man returns water to the aquifer, the process is termed artificial recharge.

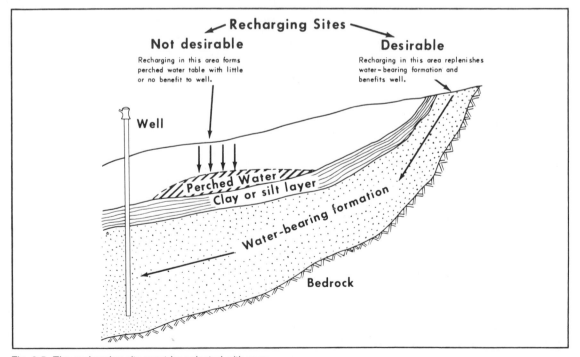

Fig. 2-5. The recharging site must be selected with care.

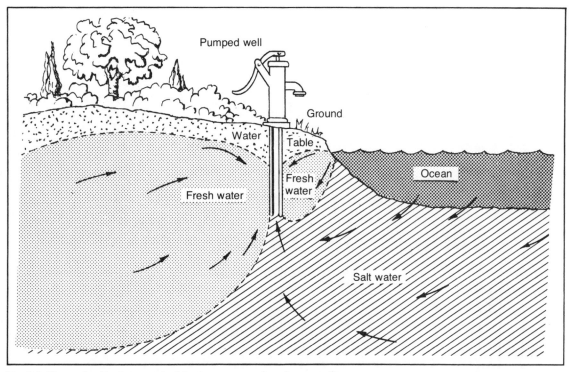

Fig. 2-6. How salt water in a nearby lake or ocean can invade an aquifer.

water is fed to a second well and not used directly can be that the creek water needs treatment to make it potable. Flowing 50 feet through the aquifer purifies (in most cases) the creek water. Another reason might be that it is less expensive and more convenient to use the aquifer as a storage tank than it would be to construct an unsightly above-ground tank that might need to be insulated against frost or to go to greater expense and labor to install an underground tank.

Remember that the second well can hold much more water than the volume of its pipe alone. If you can draw, for example, 5 gpm from well number one, you can pour just about the same quantity of water into well number two. The aquifer will soak it up.

Salt Water Intrusion. When a well is rela-

tively close to a body of salt water and the well is pumped excessively, it is possible for salt water to find its way into the well. Fresh water is lighter than salt water. If the two are not mixed they stratify; the lighter fresh water forms a layer atop the salt water. When the fresh water is pumped out, the salt water level rises. See Fig. 2-6.

When a well is located within reach of salt water, it is advisable to keep close tabs on the saline content of the water to forestall overpumping. In some instances, it may also be beneficial to artificially recharge the well in order to lower the level of the salt water in the aquifer.

For more information on the subject write to the U.S. Department of Agriculture and request leaflet No. 452, *Replenishing Underground Water Supplies on the Farm.*

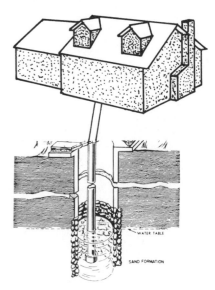

Chapter 3
Springs and Seeps

GROUNDWATER THAT ISSUES FROM THE earth either by the force of gravity or artesian pressure is classified as a *spring*. The flow might be steady or it might be seasonal. It is still a spring.

When groundwater comes to the surface of the earth over an area and there is no distinct flow—the ground is merely wet much of the time—that area is termed a *seep*.

Both springs and seeps can be developed as water-supply sources. In doing so, both require care to forestall contamination—particularly the seeps. The danger stems in part from the possible close proximity of the spring and seep water to the surface of the earth. In certain areas, both a spring and a seep are likely to drain a sink hole. The chance of this is much greater with a seep.

Described another way, there might be a depression in the earth a short distance away but higher than the seep. Rain water collects in the depression and finds it way to the seep. The problem here is that the water hasn't traveled a sufficient distance for the earth to perform its wonderous purification process. Pathogens resulting

from decayed animal matter and feces deposited in the depression can find their way into the seep.

A second problem is flooding. A heavy rain can wash over the seep and bring contamination along with it.

Your first step in developing a spring or a sink is to request that your local health officer to make a sanitary study of the area.

Protecting the Spring or Seep. Because the water for both a spring and a seep can originate close to the surface of the earth, the area around the water supply should be fenced off for a distance of not less than 60 feet. No pesticides or similar chemicals should be used in this area. Dams or channels should be provided to guide water runoff around the spring or seep area. These guides should be at least 30 feet away from the water supply. See Fig. 3-1.

DEVELOPING A SEEP

Figure 3-2 shows one way of securing more water from a seep than comes to the surface of its own

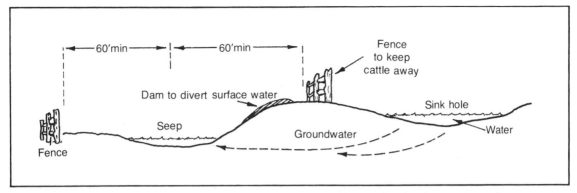

Fig. 3-1. Recommended minimum spacing and dam to protect a seep from cattle and surface water runoff.

accord. Knowing that aquifers generally follow the slope of the earth's surface and that groundwater generally runs downhill in an aquifer, you can cut a trench into the aquifer and across the groundwater's direction of flow.

The length of the trench and the depth of the trench will depend upon the quantity of water you encounter. If the seep is going to be at all useful, you are going to need a high-powered portable pump to pump the ditch dry. The quantity of water the pump discharges will give you a rough indication of the flow rate you can expect from the seep, assuming that the flow is not seasonal and you are not working during the wet season.

If there is very little water, you might try digging more deeply or extending the length of the trench. If there is obviously insufficient water, there is nothing to do but to put the earth back and look elsewhere for your water.

Assuming there is plenty of water, enlarge the center area of the trench and install a concrete storage tank. The tank should be large enough to hold several days' supply of water. The tank is provided with two wrought-iron pipes at the downhill side. One pipe handles the overflow and the other leads to the point of consumption.

In addition, you have to install two lengths of perforated 5-inch pipe to either side of the tank. These pipes lie on a bed of crushed stone and pitch upwards. Groundwater entering the trench flows through the gravel, into the pipes, and into the tank. The tank must be sealed. To increase the flow of water into the gravel and the pipes, a sheet of heavy

plastic is positioned as shown (in Fig. 3-2). With all of this done, the trench can be refilled with earth and the tank is covered.

DEVELOPING A SPRING

While each spring will require individual attention, the principles of development and protection are essentially the same. When possible, the emergence of water from the aquifer should be increased. This can be accomplished by increasing the area of acquifer exposed, removing obstructions, and enlarging the opening cracks and seams in the aquifer.

The spring must be protected from surface water—rain and flooding. The aquifer is protected from the elements, animals, and casual watertaking by a concrete barrier. In most instances, the barrier is also a tank that serves as a primary means of water storage. In some instances, water is drawn from the tank by gravity; in other cases it is pumped out. In addition, there is always a second pipe to prevent overflow.

Aquifer on Bedrock. Figure 3-3 shows a developed spring that receives water issued from an aquifer that rests on bedrock. The aquifer is blocked and held in place with large stones laid up dry (no cement). The concrete tank rests in part upon these stones and in part upon the bedrock.

In constructing this tank the first step would be to shelve the hillside. The second step would be to erect the coarse stone barrier. Then forms could be constructed for the tank. The bottom of the tank can be kept dry with the aid of a high-power portable

25

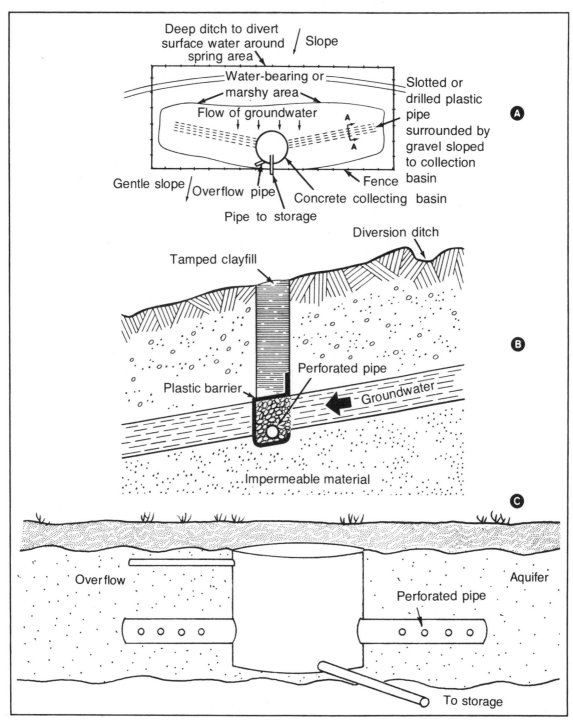

Fig. 3-2. A: Slotted or drilled plastic pipe can be used to collect near-surface groundwater. B: Profile of A-A section in upper drawing. C: A concrete tank and perforated pipe can be used to collect groundwater from a near-surface aquifer.

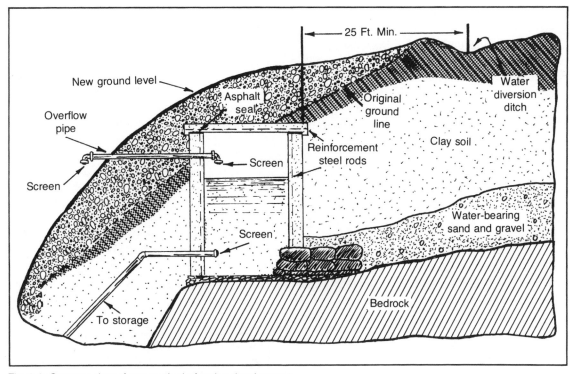

Fig. 3-3. Cutaway view of one method of spring development.

water pump. After the tank has been completed and the pipes are installed, it is covered with a layer of earth for frost protection.

Sub-Soil Aquifer. The tank used in conjunction with the spring shown in Fig. 3-4 could be made from sections of concrete pipe lowered into place. To accommodate the rock base at the bottom of the tank, a portion of one section of pipe could be cut off. The major problem in a setup such as this is keeping it dry while you are excavating. This can

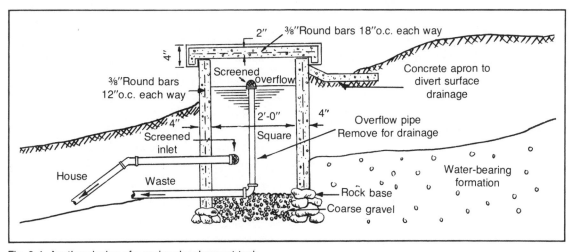

Fig. 3-4. Another design of a spring-development tank.

only be done with a high-powered pump going constantly.

Unconsolidated Rock Aquifer. In this situation, the groundwater issues from cracks in unconsolidated bedrock. These cracks or faults should be enlarged. Try prying with crowbars and the like. The weathered and loose rocks near the fissures should be removed, and all cracks and fissures that will not lead into the tank should be grouted (sealed with concrete).

Figure 3-5 shows a number of pipe sections being used to form the tank. The exterior of the tank so formed is given a watertight coating of cement about ½ inch thick. This can be made by mixing 2 parts sand with 1 part mortar cement and adding a quantity of Anti-Hydros. A watertight or waterproof exterior layer is not necessary with the previously described designs because the level of water within the tank is or will be as high as the level of water outside the tank and in the zone of aeration. Thus there will be little water movement from the zone of aeration into the tank.

Exposed Rock Aquifer. In hilly areas, you will often see water seeping from cracks in the sides of exposed rocks. Sometimes the seepage can be increased to a useful flow rate by enlarging the cracks and crevices and sometimes by driving pipes deeply into the cracks. Sometimes cutting back into the rock or exposing more of the rock and its cracks and crevices helps. When the rate of water flow is satisfactory, a concrete storage tank can be erected around the spring (as shown in Fig. 3-6).

If there is no danger of frost, the tank can be left exposed. If there is a chance of frost, the tank should first be sealed with a layer of mortar cement, as previously described, or with a layer of asphalt and then covered with soil.

Surface Springs. The technique used to de-

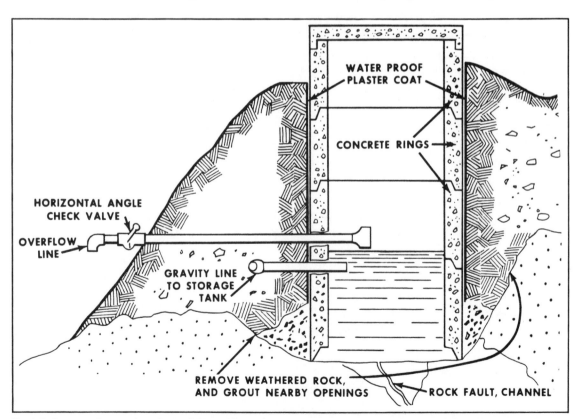

Fig. 3-5. Concrete rings such as those often used for storm sewers can be used to construct a tank for spring development.

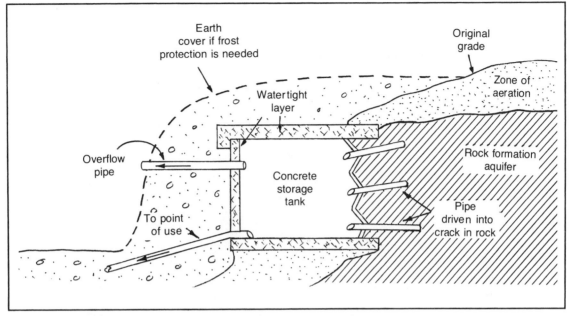

Fig. 3-6. Underground water is seeping from the face of an exposed rock formation.

velop a spring that bubbles to the surface of a relatively level piece of ground is similar to that used to develop or construct a dug well. The difference, of course, is that unless you have a high-powered pump going all the time you will be working in mud and water. See Fig. 3-7.

Simply dig a large hole directly down into the spring, keeping the pump going all the time. The diameter of the hole should be slightly larger than the outside diameter of the tank you are going to position. This can be a section of clay or concrete pipe. The depth of the hole will depend upon the

overall height of your tank or pipe and how far below the surface you want it to be.

With the hole dug, cover the bottom with 6 inches or so of coarse gravel. Level the surface of the gravel and then sit the pipe directly on the gravel. The cover can be made by pouring concrete into a wood form and reinforcing the concrete with heavy-gauge wire.

The two pipes entering the tank warrant mention. The service pipe, the pipe from which you will draw your water, is treated conventionally; you simply connect it to your water system or tank. The

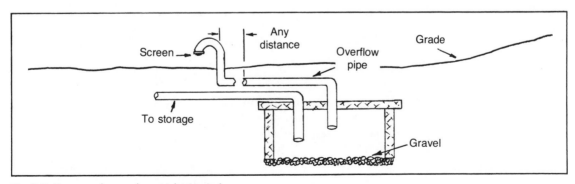

Fig. 3-7. How a surface spring can be treated.

Table 3-1. Typical Hydraulic Ram Specifications.

RAM NO.	INTAKE (drive) PIPE SIZE	DISCHARGE (delivery) PIPE SIZE	INTAKE CAPACITY (G.P.M. used) Min. Max.	MINIMUM VERTICAL FALL REQUIRED (in feet)	SHIPPING WEIGHT IN POUNDS (approx.)
RIFE "NEW MODEL" SERIES BU RAMS (4-bolt Design) Maximum vertical fall 15 ft. Maximum vertical lift 150 ft. Unit includes strainer for intake end of drivepipe					
10 BU	1-¼"	¾"	2 - 7	3	120
15 BU	1-½"	¾"	5- 13	3	120
20 BU	2"	1"	8 - 20	3	120
RIFE "EVERLASTING" STANDARD RAMS A more rugged development of the previous series "A" 6-Bolt Design Maximum vertical fall 25 ft. Maximum vertical lift 250 ft. Unit includes strainer for intake end of the drivepipe					
10 SU	1-¼"	¾"	3 - 10	3	125
15 SU	1-½"	¾"	5 - 14	3	125
20 SU	2"	1"	10 - 22	4	125
20 SUL	2"	1"	12 - 30	4	270
25 SU	2-½"	1"	15 - 45	4	270
30 SU	3"	1-¼"	20 - 70	4	270
40 SU	4"	2"	35 -125	4	565
60 SU	6"	3"	75 -350	4	1325
RIFE "UNIVERSAL" HEAVY DUTY RAMS (6-bolt Design) Maximum vertical fall 50 ft. Maximum vertical lift 500 ft. Unit includes strainer for intake end of drivepipe					
10 HDU	1-¼"	¾"	3 - 10	3	170
15 HDU	1-½"	¾"	5 - 15	3	170
20 HDU	2"	1"	10 - 25	4	170
20 HDUL	2"	1"	12 - 33	4	310
25 HDU	2-½"	1"	15 - 45	4	310
30 HDU	3"	1-¼"	25 - 75	4	310
40 HDU	4"	2"	35 -150	5	565
60 HDU	6"	3"	75 -400	5	1325

Courtesy Rife Hydraulic Engine Mfg. Co.

overflow pipe must have its end covered with screening to prevent the entrance of soil, and it must terminate a foot or so higher than the surface of the original earth at the spring. That is no problem. In frost country, you might have to carry this pipe underground until it can be brought up inside the house. The pipe can be small diameter plastic. It doesn't need to be any where as large in diameter as the supply pipe. If you do not provide an overflow pipe, and the service line is closed, the water in the tank will attempt to reach its original, above-ground level. In doing so, it will lift the tank cover.

HYDRAULIC RAMS

Hydraulic rams (Table 3-1) are self-powered water pumps that are almost ideal when you have water to "burn." Very simply they are impulse pumps (Fig. 3-8) that are powered by the pressure of the water that enters the pump. No external power, motor, engine or wind-driven fan is required. The water

Fig. 3-8. An impulse pump. Photo courtesy Rife Hydraulic Engine Mfg. Co.

entering the pump does it all. Nothing is free. Much of the water that enters the pump is lost, wasted.

To use an impulse pump, you need a physical layout wherein you can position the pump below the spring so that the water can flow down into the pump, and you must have some means of disposing of the waste water.

Typically, the impulse pump or ram is positioned on a hillside below the spring. See Fig. 3-9. Water flows down into the ram, the ram pumps a

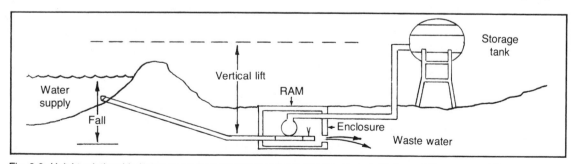

Fig. 3-9. Height relationship between an impulse pump (ram) water source and discharge. The energy of the waste water drives the pump.

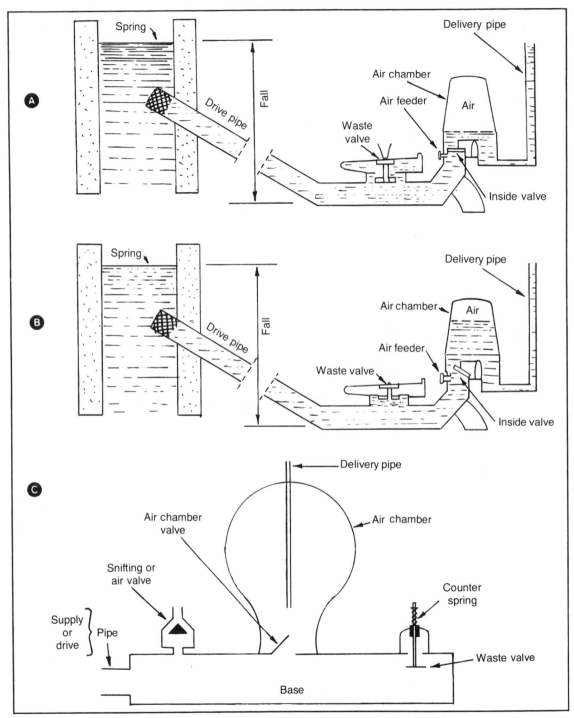

Fig. 3-10. The operating cycle of an impulse pump or a hydraulic ram: A shows the waste valve in its open position; B shows this valve closed and water entering the air chamber; C shows the essential parts.

portion of the water back up the hill (or wherever it will be stored), the balance of the water, the waste, is discharged and runs down the hill and is used for irrigation or simply permitted to soak into the earth.

How it Works. Water flows down the drive pipe and into the air chamber where it compresses the air. The flow of water stops because air pressure is now equal to water pressure. The compressed air now pushes a smaller quantity of water through the delivery pipe. More water now flows into the air chamber and the cycle is repeated. This can go on at the rate of 25 to 100 times a minute. See Fig. 3-10.

There are three valves in the system. The waste valve is located in the supply or drive side of the pump. This closes when water enters the pump. Another valve is in the air chamber. It is open when water enters, but closes when the chamber is as full as it is going to get. The third valve is called the air snifter. It is positioned after the waste valve to admit a little air between strokes so that the air chamber is always full of air.

Efficiency. In terms of water received to water delivered, the efficiency of hydraulic rams is very low. Under ideal conditions, water output does not exceed 25 percent of the water entering the pump. Under extreme conditions, less than 10 percent of the water will be pumped to its desired destination. To approximate the quantity of water discharged, use the following formula:

$$\frac{\text{Fall in feet} \times \text{gallons per minute supplied} \times 40}{\text{feet water is to be elevated}}$$
$$= \text{gallons delivered per hour}$$

If the fall is 6 feet and the ram is supplied with 30 gallons of water a minute, you have 6 × 30 = 180. This is multiplied by 40, which gives 7200. If you want to raise the water 40 feet, divide 7200 by 50, which gives 144 gallons of water delivered per hour.

Note that this formula does not take into consideration pipe friction (which would lower the output). Note also that this ram is supplied with 1800 gallons of water an hour and receiving only 144 in return. But, and it is a big but, we do not have to

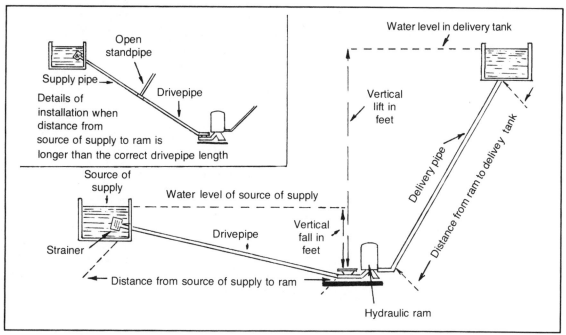

Fig. 3-11. Data required to determine proper ram pump size and its limits of performance. Courtesy Rife Hydraulic Engine Mfg. Co.

supply any power to the ram. The falling water does it all. As you can see from the formula, the higher you want to drive the water, the less you are going to get.

Figure 3-11 shows the major points of hydraulic ram installation. The ram must be at least 3 feet below the spring or brook. The ram must be protected from leaves and dirt and a frost-proof enclosure where frost is a problem. Waste water from the ram must flow freely away. Water above the snifter will stop it from operating. The intake or drive pipe should be no less than five times longer and no greater than 10 times the height of the fall. The discharge and intake pipe should match the required pipe diameter on the pump, and there should be no diameter change in either of these two pipes. The intake or drive pipe must be screened to prevent the entrance of leaves and the like. The ram must be bolted in place.

Ram Size. To determine ram size required, measure the normal flow of water from the spring in gallons per minute. Measure the drop and the desired rise. Measure the lengths of the piping that will be required. Make a sketch of your proposed layout. Compare your needs to the manufacturer's literature. Choose the pump that will provide the flow rate you want given the available drop and incoming water quantity.

Operations. Open the gate valve in the supply pipe and depress the waste valve. This permits water to enter the ram. You might have to repeat this a few times. In other words, you want to get the ram "vibrating" or stroking by itself. If the pump will not stroke of itself, try varying the adjustment on the waste valve. Try for a speed that appears to be smooth and steady and close to what the manufacturer recommends.

Troubleshooting. If the ram stops with the waste valve in the open position, there is an air leak either in or near the valve or in the pipe connections. If the ram stops with the waste valve in the down position, chances are that the quantity of water reaching the pump is insufficient. Sometimes adjusting the valve to a shorter stroke will enable it to continue with a less-than-normal water supply. As an alternative, the valve in the output line can be partially closed, thus drawing less water from the pump.

If the ram works but delivers no water, the spring might have run dry, there might be an air leak in the supply pipe, or water might have filled the air chamber because the snifter valve is stuck. Close the input pipe line, drain the pump, and clean or replace the snifter.

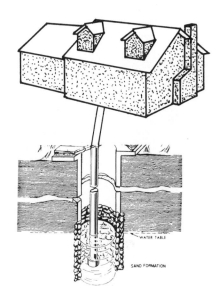

Chapter 4

Finding Water

WATER TABLE

SAND FORMATION

APPROXIMATELY 90 PERCENT OF ALL THE WATER to be found beneath the surface of the earth in this country is found within the top 200 feet. The average depth of all residential water wells is about 50 feet. While these figures might be comforting, they do not assure you of finding water at a depth of 50 feet or at worst at 200 feet. There are areas where the water lies at tremendous depths—thousands of feet. How do you find where the shallow wells can be dug or drilled? How does one avoid drilling 200 and more feet to find usable quantities of water?

GEOLOGY AND GEOGRAPHY

The depth at which you will find an aquifer and the quality and the quantity of water an aquifer can supply will depend upon its location. The greater the total annual rainfall in an area, the greater the likelihood of water being found in copious quantities relatively near the surface of the earth. The western part of the United States is relatively arid—with an annual rainfall of less than 20 inches—while the eastern part of the United States is fairly dripping with rain and is classified as humid. The average annual rainfall in the East is over 30 inches. As is shown in Fig. 4-1, the average annual rainfall in the continental United States varies from the desert, with less than 10 inches, to heavy rain areas, with more than 60 inches per year.

But rain alone does not make for successful water wells. The geology of the soil and subsoil might be such that there might be aquifers present and exposed to rainfall. But all heavy rainfall areas do not lie above good aquifers. It takes a combination of rain and a suitable underground layer of sand, gravel, or unconsolidated rock to provide a well with water. Together, geology and geography can be termed a primary indicator of the chances of striking water.

Relative Elevations. Groundwater generally tends to follow the surface of the earth when and where the surface runs downhill. Your chances of finding water by digging your well at the top of a hill is far less than your chance of digging a well and

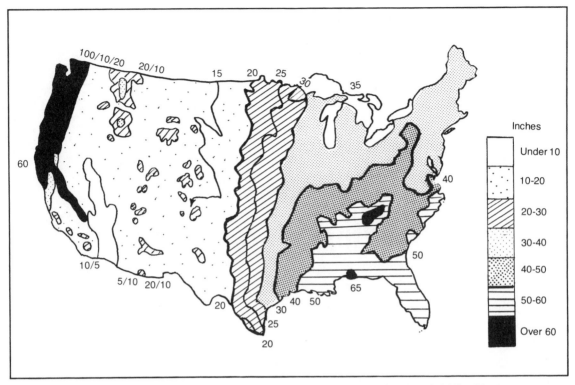

Fig. 4-1. Average annual rainfall in the United States. Areas in the United States capable of yielding 50 gpm or more.

finding water at the bottom of a hill. This is not always the case because underground water can travel for hundreds of miles.

Topographic maps are useful for locating wells. A topographic map indicates the relative elevations of the land and usually also indicates rivers, streams, lakes, and the like. Such a map of your area will quickly tell you how high your property stands in relation to the rest of the world and will help you to evaluate your chance of striking water.

The U.S. Geological Survey Topographic Maps carry all the above information and more. They can be purchased for a few dollars and are available for every area in the United States. Write to: U.S. Department of Interior, Geological Survey, 1200 South Eads St., Arlington, VA 22202.

Local Water Indicators. In addition to elevation and proximity to large bodies of water, there are other guides to locating suitable areas where wells can be dug successfully.

Look for water seepage (constant wet spots indicating springs and seeps). You either develop the spring or seep or drill your well into the same aquifer. Position the well on ground higher than the spring or seep.

Examine whatever bedrock protrudes from the soil. Hard rock such as granite indicates very little chance of an aquifer. Your well has to intersect cracks in the rock to find water. The presence of soft stones such as limestone, sandstone and the like indicates a good possibility of finding water at 50 feet or less.

Plant Indicators. There is a group of plants called phreatophytes that can exist only where their root systems can reach the water table. The word *phreatophyte* comes from two Greek words meaning well plant. Not only do these plants indicate groundwater, but to a certain extent they can tell you how much water is present and at what depth.

A willow or a cottonwood tree growing in a normally dry region indicates considerable water within some 20 feet of the surface. A good-sized tree can transpire several hundreds of gallons of water a day. You can be reasonably certain there is good water in a generous quantity present in the earth when you see these trees. Some species of birch, sycamore, bay, live oak, alder and red oak also indicate groundwater at a fairly shallow depth.

When and where these trees are found in humid regions, however, their value as indicators are greatly reduced. There might be so much water in the zone of aeration that the plants do not have to reach deeply into the earth to survive. Each region of our country has its own group of phreatophytes. Ask your local farm advisor or agricultural experimentation station for their names. Figures 4-2 through 4-10 illustrate the more common phreatophytes found in arid areas of this country.

DOWSING FOR WATER

Some people claim to have been born with the ability to find underground water with the aid of a divining rod. The rod is most often a forked twig

Fig. 4-3. Cane and reeds—good quality water 10 feet or less below the surface.

Fig. 4-2. Sedges, cattails and rushes—good water near or just below the surface.

Fig. 4-4. Salt bush, grows along the margins of salt flat. Water is present but its quality has to be checked.

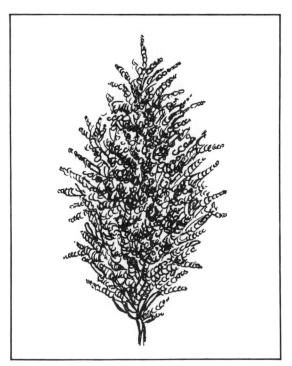

Fig. 4-5. Pickleweed—near surface mineralized water in soil with 1 to 2 percent salt content.

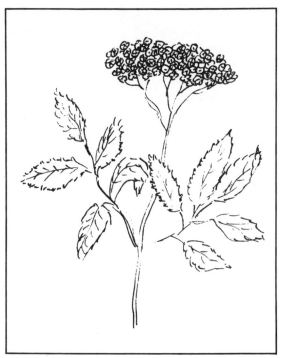

Fig. 4-7. Small trees and elderberry shrubs—water within 10 feet of the surface.

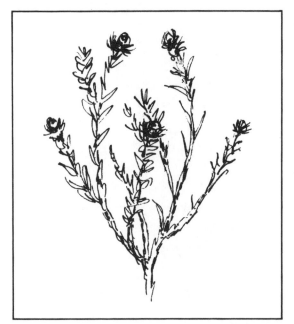

Fig. 4-6. Arrow weed—good-quality water 10 to 20 feet below the surface.

Fig. 4-8. Rabbit brush—water within 15 feet of the surface.

Fig. 4-9. Black grease wood—mineralized water 10 to 40 feet down.

(Fig. 4-11) cut from a peach, willow, hazel, or witch hazel tree or bush. But sometimes the rod is no more than a straight twig. Sometimes it is even made of metal or something as common as a buggy whip.

The usual technique of holding the twig has the diviner's hands on the two portions of the fork, with the butt end pointing away from the diviner. Some diviners hold the twig with the butt toward their body. Those operating with a straight twig hold it by its thin end in a horizontal position.

Upon passing over underground water, the forked twig's end will be attracted downward. Some diviners state that the twig will actually spring around in their hands. When a straight twig is passed over a source of underground water, the twig will bend downward. The distance the twig bends and the strength of its response serves to indicate the quantity and the depth of the water to be found directly beneath the divining rod.

Does It Work? You can, if you prefer, cut yourself a forked twig (Fig. 4-12), hold it gently in your hands, and walk about your property until you secure a response from the twig. The plants useful for divining rods require a lot of water. The theory is that the twigs from those trees are attracted to water. A tree's roots will grow in the direction of water; they will grow more or less sideways if that is where the water is located.

Should the twig fail to respond in your hands, diviners will tell you that you have not been born

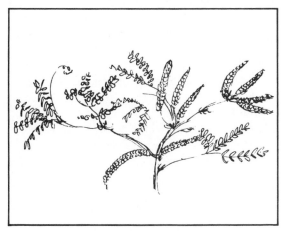

Fig. 4-10. Mesquite—good water 1 to 50 feet down.

Fig. 4-11. The old "forked stick" technique.

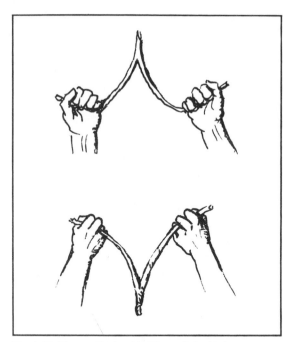

Fig. 4-12. Two methods of holding a forked twig.

with the gift of divination or you are not holding the twig properly. Holding it too loosely, they say, is as bad as holding it too tightly.

Should the twig respond vigorously above one spot on your property, have you truely found water? Should you dig or drill there? Because there is water to be found at some depth below the surface of almost all of continental United States and because considerable water exists at depths of 50 feet or so in much of this country, chances are fair that you have found water. On the other hand, the United States Geological Survey states categorically that water witching is practically useless. They state that a tremendous amount of investigation has been bestowed on the subject with an absolute lack of positive results.

Origin. The origin of the divining rod has been lost in history. Ancient writings indicate that "magic" wands of one kind or another were used by ancient seers to find lost objects, forecast events, and find water. The divining rod is mentioned several times in the Bible, tying it in with wonderous performances—particularly in the book of Moses.

The Persians, Scythians, and the Medes used them. Marco Polo encountered the use of rods for divination in China. The ancient Germans, according to Tacitus—an early Roman historian—favored branches from fruit trees for dowsing.

For more information on the interesting subject of dowsing, write to the Government Printing Office, Washington, D.C., and ask for the *Water-Supply Paper 416, The Divining Rod* by Arthur J. Ellis.

Modern Techniques. Professional well diggers use electrical resistivity measurement and seismic refraction (studies of induced vibrations in the earth) to help them determine the nature of the subsoil. They also use gamma-ray logging; this technique measures the natural radiation in a borehole.

OTHER SOURCES OF INFORMATION

The following sources of information might not be available or you might not be able to secure their cooperation.

Neighbors. If there are operating wells within a quarter mile or even a half mile, depending upon the local topography, it is well worth your time to discuss them with their owners. Learn how deep they are, how much water they yield, the quality of the water, the nature of the aquifer and whatever else they know. If there are a number of wells and you can secure the aforementioned data, you can draw a sketch of the probable shape and depth of the aquifer. While it is possible to drill and miss an aquifer altogether, aquifers are generally hundreds of yards wide in each direction. If you drill between two producing wells, chances are you are going to hit water at the same depth or close to it.

Local Well Diggers. If you are planning to dig your own well, the local well diggers—you can find them in local telephone directories—might be reluctant to impart their information to you. Perhaps they will give you their data for a fee. In any case, well diggers keep logs of well depth, the nature of the aquifer, the level at which it occurs, the type and diameter of the well drilled, and the yield obtained. This is valuable information and it is well worth buying.

Commercial well diggers will also drill a test

well for a fee. In some instances, it might be worth your money to invest in a test boring to learn beforehand what you will encounter in the way of rock, and the depth you will have to go to find water. Knowing your problem might save you from bringing in the wrong equipment or wasting time with methods unsuited to the subsoil you have to penetrate.

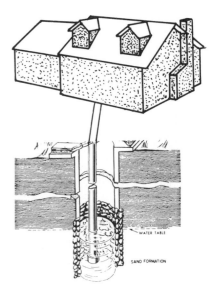

Chapter 5

Before You Dig or Drill

YOU HAVE FOUND THE SPOT FOR YOUR WELL. All the indicators point to groundwater at the location you have chosen. Now it would seem there is nothing more to do but dig or drill into the earth, but that is not the case. To simply go ahead believing that all you risk is not finding water is a serious error. Before you spend time, money, and efforts digging your well or drilling it, you *must* be as certain as you can that the water you draw will not be contaminated (Fig. 5-1). This is not always easy to do, but there are steps you can take to protect yourself before you dig or drill your well.

The first step is to position your well as far as you can or as far as practical from obvious sources of contamination. There is no positive safe distance from any contamination. Groundwater can travel a great distance, and if contamination seeps into the groundwater it can be carried to your well.

Further precautions consist of positioning the well, if at all possible, uphill from the sources of contamination. Also, position the well beyond the building's roof drip line and away from dry wells.

Great danger can exist in distant and otherwise hidden sources of contamination. Most of us have heard of the Love Canal in New York and how the toxic wastes dumped into the canal many years ago have leached through the earth to poison the surrounding area. The Love Canal is not the only toxic waste dump; there are thousands of them.

For a long time, people believed in the "magic" of earth filtration and the work of its microbes. While microbes and the earth still do a magnificent purifying and filtering job, microbes are not effective against nonorganic substances and the earth is a filter with a finite capacity. Sooner or later it loses its ability to filter, assuming, of course, that the poisonous chemicals have to travel some distance through the earth and do not move on through the wide spaces found in coarse sand and gravel.

Thousands of Poisons. There are more than 63,000 synthetic, organic chemicals now being marketed, few of which can be handled by microbes. All of them are potentially lethal or at least harmful when ingested even in minute quantities. In addition to the esoteric poisons, there are toxic lead, mercury, TCE (Trichloroethylene, a common

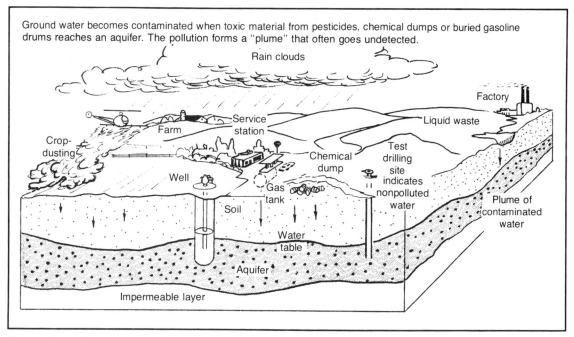

Ground water becomes contaminated when toxic material from pesticides, chemical dumps or buried gasoline drums reaches an aquifer. The pollution forms a "plume" that often goes undetected.

Fig. 5-1. How pollution can reach an aquifer.

dry cleaning agent) benzene, toulene, carcinogenic vinyl chloride, gasoline, carbon tetrachloride, aldicarb and DBCP (both pesticides), to list just a few.

Thousands of Toxic Dumps. There are 51,000 known toxic dumps in this country. Environmental Protection Agency officials estimate that three-fourths of the abandoned and still-in-use dumps are leaking their poisons into the ground. Once an aquifer is contaminated, there is no known practical way of cleaning it up. In some areas, it takes very little to contaminate an aquifer and make the drinking water dangerous. In Florida, where in some areas the water table is only 6 inches below the surface, runoff from a car wash can cause trouble.

A Wide Range of Illnesses. The physical and mental difficulties produced by tainted water range all the way from cancer to constant headaches, acute kidney problems, malformed births, nausea, brain tumors, and more. Former EPA official Eckhardt Beck says, "The contamination of groundwater is the environmental horror story of the '80s."

Millions Affected. Although no more than 2 percent of the U.S. population of the United States is presently affected by contaminated drinking water, their total number is easily 5 million. Over the nation, more than 2000 wells have been shut down. The Federal agency assigned to the task is understaffed and underfinanced. There is no standard of purity for wells supplying fewer than 25 individuals. In Michigan, there are 200 crucial sites that must be investigated, but the Michigan agency has only one drill suitable for the purpose. With one drill, they cannot examine and inspect more than 15 dumps a year.

Half a million new wells are dug every year. There is 50 times more groundwater in this country than fresh surface water.

Poisoned water is difficult to detect. You cannot tell it is bad until it becomes so contaminated that you can taste the poison. Until that point, the water might be clear, cold, and sparkling clean.

Distance from a toxic dump is no guarantee that your well will not be affected. Toxic waste first dumped on a regular basis into a pit 90 miles outside of Memphis in 1964, reached the water system of the city in the late '70s. Citizens became ill upon drinking their tap water and even passed out in the shower.

Look Before You Dig. The best way to protect yourself and family against contaminated drinking water is to check the proposed well site carefully before you begin any well work. Check for possible sources of surface contamination: septic tanks, drain fields, barns and the like.

Obvious sources of contamination include nearby gas stations, whose underground tanks might be seeping gasoline into the earth. Check with your local health department. They should know the sanitary condition of the aquifers beneath your property.

If you have any doubts, go the expense of a test well to sample the water. Have the water tested. This is neither simple nor inexpensive. The range of possible tests can run to a thousand dollars or more. If at all possible, try to learn through your neighbors and the local health department what the lab should look for. The cost will be far lower.

In any case, don't guess at the quality of the water that comes up. *Don't* depend upon your sensitive palate or sense of smell. Poisoned water is insidious; it often takes a long time to kill.

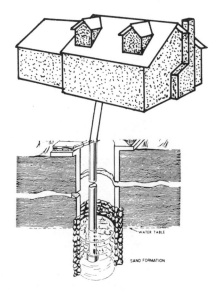

Chapter 6
Driven Wells

WHEN GEOLOGICAL CONDITIONS ARE RIGHT, driven wells are the easiest, fastest and simplest of all types of wells to build. Essentially, the driven well is just that. It is a pipe, tipped with a combination well screen and point, that is driven into the ground with either hand tools or powered tools.

Pipe in diameters from 1¼ to 4 inches can be driven this way to depths of up to 50 or more feet into the ground. The quantity of water resulting depends upon the aquifer the well point penetrates, the water admittance area presented by the screen (fine holes less water, large holes more sand), and how well the well is developed. See Fig. 6-1.

The driven well is usually the first choice for a summer cottage or a camp site where no water is otherwise readily available and you do not want to spend a lot of money or time on a water supply. In the winter, when the well is not used, the pump can be removed and the well can be capped. Frost will not damage the driven well when this is done because the water level in the well will drop to the level of the water table, which is usually well below the frost line. In any case, water freezing in the well has plenty of room to expand upward without causing damage.

Driven wells are also used for year-round water supplies. Where frost is a problem, the well pipe is terminated below the frost line and connected to the pump by means of a fitting called a *pitless connector.*

If and when a single driven well will not produce the desired quantity of water, a number of wells can be driven and their outputs can be connected together. Essentially, the driven well itself consists of only two parts: the well point and the pipe that is connected to it. See Fig. 6-2.

Well Points. The well point is an especially constructed screen provided with a reinforced steel or bronze tip at one end and pipe thread at its other end. Points are selected with screen openings suitable to the aquifer where the point will enter. For coarse gravel, a screen with wide openings is chosen. For fine sand, a screen with very fine openings

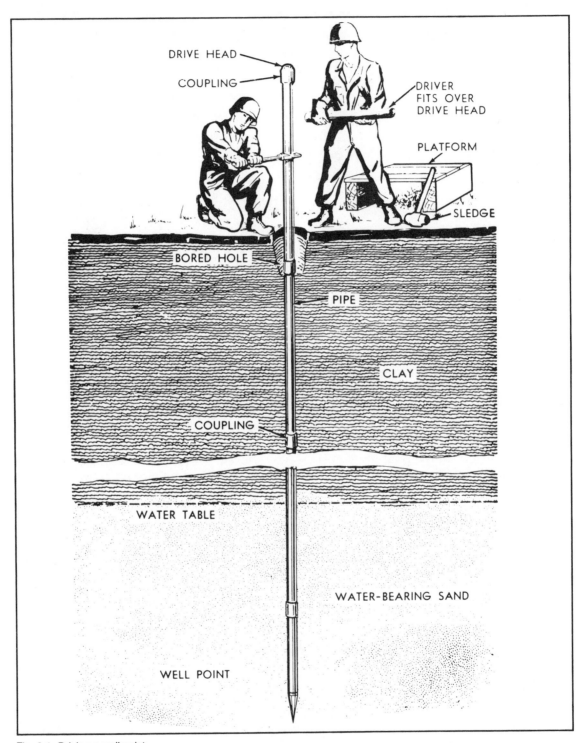

Fig. 6-1. Driving a well point.

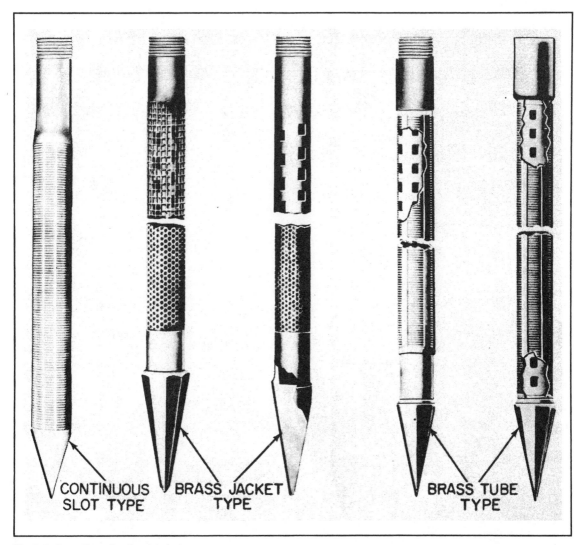

CONTINUOUS SLOT TYPE　　BRASS JACKET TYPE　　BRASS TUBE TYPE

Fig. 6-2. A number of well-point designs.

is chosen. It is a compromise between reducing the inflow of water and admitting sand. See Fig. 6-3 and Table 6-1.

Pump Type. The choice of pump type will depend upon the depth at which the water table exists and how far the water table will drop within the well pipe when pumping commences, and, of course, the ID of the pipe being used.

If the water level in the well pipe will not drop below 20 feet or so and the well is at sea level, a suction-type pump will be satisfactory. This can be

motor driven or it can be a standard, regular, hand-operated pump. Above sea level, the 20-foot operating depth will be reduced proportionally for vacuum or suction pumps. This figure might seem low in view of the ability of air pressure at sea level to support a 32-foot column of water, but no pump is perfect and air leaks hold the operating depth down.

When the water level in the well is below 20 feet or will drop below this distance, you can use a special hand-operated pump or a motor-driven

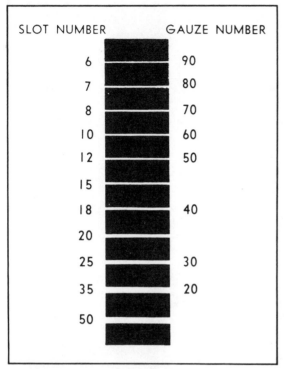

SLOT NUMBER	GAUZE NUMBER
6	90
7	80
8	70
10	60
12	50
15	
18	40
20	
25	30
35	20
50	

Fig. 6-3. Standard slot numbers and equivalent gauze numbers. The white spaces represent the spaces to be found in a well screen with that number.

pump. In the latter case, you must use a pump or pump system that will fit into your well casing. See Fig. 6-4.

On the plus side, the driven well is very simple, comparatively inexpensive, and can be quickly installed with a few hand tools and simple muscle power. In suitable soil, hand power alone can drive the well-point down 30 feet or so. With power assist, a depth of 50 and more feet are easily reached, soil conditions permitting.

The one rub is soil conditions permitting. If the well point encounters even a small stone—say 6 inches across—head-on, driving will be stopped. The well point and attached pipe has to be pulled and the stone must be removed, if possible. If there are a lot of stones and if the stones are mixed with clay, driving becomes very difficult—if not impossible. Pound too hard on the top of the pipe and the well point might be crushed and the pipe might buckle.

Well-point screen areas are comparatively small. If the aquifer the point enters does not have a strong flow, the water that can be drawn from the well will be limited. One possible solution is to try driving the well point more deeply into the earth in hopes of reaching a stratum having better water flow. A coarser aquifer might provide this.

Another possible solution might be to pull the well point and have a go elsewhere. Another might be to drive a second and a third well and couple them together. And another would be to try a well point that is longer or has a coarser screen in order to admit more water. Generally, going to a larger-

Table 6-1. Steel-Wound, Open-End Well Screens.

Pipe Length Wrapping Length Inches	I.D. Pipe Size Inches
30×24	
36×30	
48×42	
60×54	1¼
72×66	
96×90	
120×114	
48×42	
60×54	1½
120×114	
36×30	
48×42	
60×54	2
72×66	
96×90	
120×114	
60×54	
72×66	2½
96×90	
120×114	
60×54	3
72×66	
120×144	
48×36	
60×48	
72×60	4
96×84	
120×108	

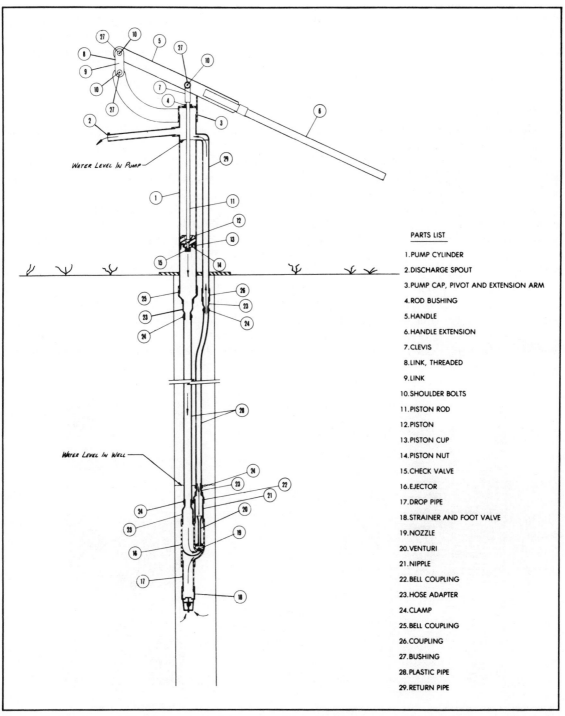

PARTS LIST

1. PUMP CYLINDER
2. DISCHARGE SPOUT
3. PUMP CAP, PIVOT AND EXTENSION ARM
4. ROD BUSHING
5. HANDLE
6. HANDLE EXTENSION
7. CLEVIS
8. LINK, THREADED
9. LINK
10. SHOULDER BOLTS
11. PISTON ROD
12. PISTON
13. PISTON CUP
14. PISTON NUT
15. CHECK VALVE
16. EJECTOR
17. DROP PIPE
18. STRAINER AND FOOT VALVE
19. NOZZLE
20. VENTURI
21. NIPPLE
22. BELL COUPLING
23. HOSE ADAPTER
24. CLAMP
25. BELL COUPLING
26. COUPLING
27. BUSHING
28. PLASTIC PIPE
29. RETURN PIPE

Fig. 6-4. Design of a special hand pump that will draw water without dependance upon drawing a vacuum. Courtesy Hipps Hand-Pump Mfg. Co.

diameter pipe and a larger-diameter well point does not produce proportionally more water. Little effort should be wasted on any of these changes until the well has been developed. This is a procedure that must be followed with every new well to cleanse and clear its water and to produce maximum flow. See Chapter 12.

On the negative side is that depth is limited to a maximum of about 50 feet. Beyond this depth, the pressure on the well point and associated piping becomes so great that one or both give way. Depth limitation is a severe handicap when you are not at all certain of the water table or the stratum of aquifer you want to reach. If you drive a well point 50 feet and find no water, you cannot go deeper. You have to pull the 50 feet of pipe or abandon it. With other drilling methods you can continue on down.

A driven well consists of three major parts: The well point, the pipe to which it is attached, and the pump that draws water from the well.

Well Points. The well point is a specially constructed screen, tubular in shape, furnished with a steel or bronze point at one end and thread that engage the following pipe at the other end. Well points are manufactured in a number of styles, sizes, and with different well-screen openings. When ordering, you must specify the ID (internal diameter) of the pipe you plan to use and whether it is standard galvanized water pipe or special well-point pipe called risers. The two types of pipe have different threads and one will not properly engage the other. See Fig. 6-5. You will also want to specify the screen-opening size and describe the nature of the aquifer: gravel, coarse sand, or fine sand.

Pipe and Risers. For shallow driven wells in sand or soft earth, standard galvanized water pipe will work satisfactorily. It will, however, rust through with time. Threading the pipe removes the zinc from the pipe ends and that sets up a galvanic couple that speeds corrosion. Standard galvanized pipe usually comes in 10-foot or 20-foot lengths. Both are much too long to drive into the earth in one piece. You must cut them to about 5-foot lengths. Remember that all metal pipe is sized by its internal diameter (ID) and the 1¼-inch minimum pipe size

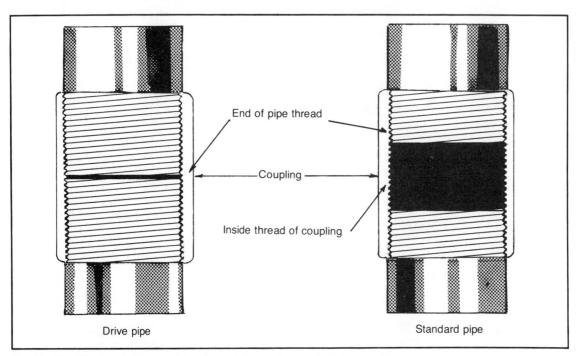

Fig. 6-5. The difference between a drive-pipe coupling and a standard pipe coupling.

recommended has an OD (outside diameter) of about 1⅝ inches.

The risers are manufactured in 5-foot and 6-foot lengths. There is usually no need to cut these steel pipes. If you do, try and see if you can secure a nipple (short, threaded piece) from your suppliers rather than cutting the riser and running up a thread on the cut end.

TERMINATION

If you are going to draw water from your well with a hand-operated pump—either of the suction, pitcher-spout type or one of the special pumps, which, although hand operated, will go down below the suction line—terminate the well pipe at any convenient height and screw your pump into place directly on the top end of the well pipe. Depending upon the pipe and the pump, you might have to use an adapter between pipe and pump.

Should there be any danger of frost, empty the pump chamber and let the water out of the pump itself (assuming some has remained and not seeped back down into the well). Then cover the pump with a tarp or a similar covering. If there is no drain valve, unscrew the pump and drain it. Then either store it out of the weather or put it back in place. If you store it, take care to seal the top of the well against rain and debris.

Apron. When the pump empties directly or almost directly above the well head, it is best to protect the water supply with a concrete apron around the pipe end (Fig. 6-6). The apron serves to keep the earth nearby from turning to mud from spilled water and it also prevents the same water, now soiled, from seeping into the aquifer alongside the pipe.

The concrete apron should be 3 or 4 feet in diameter, some 3 to 6 inches thick, and sloped upward toward the well pipe at its center. You can use a simple hoop of wood or metal as a form, pour the concrete into the hoop, and—if you use a sufficiently thick mixture—you can easily slope the concrete as required.

Underground Termination. When you need to draw water from the well through the winter and cannot remove the pump itself, or when you do not want a pump in your yard or want an automatic water supply, the well can be terminated underground. This can be done in two ways. An elbow can be connected to the top of the well pipe or the connection can be made with a pitless connector. The advantage of the pitless connector is that it permits you to easily uncover the top of the well pipe, should that be necessary. The pitless connector can be positioned above ground or beneath the earth. See Figs. 6-7, 6-8 and 6-9.

There should be no problem when you install a pumping system that depends upon a pipe or pipes that run down into the well pipe itself or a system that encompasses a pump motor near or below the water line. When you employ a straight vacuum pump of any kind, bear in mind that you will lose vacuum in the horizontal pipe line in addition to the vertical well pipe itself. Therefore you cannot depend upon a straight vaccuum pump dependably lifting water a distance of 20 feet. The total height will be reduced.

HAND TOOLS AND EQUIPMENT

You will need two large Stilson wrenches (large enough to handle the pipe or risers you plan to use), a small spirit level, pipe thread compound (pipe dope), and a length of fishing line with a small float and lead weight at one end (to determine water table level). In addition, you will need some means

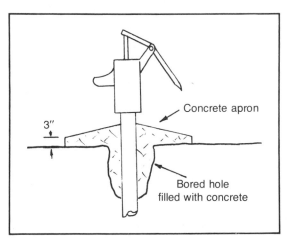

Fig. 6-6. The space around a hand pump and a bored hole can be sealed with concrete.

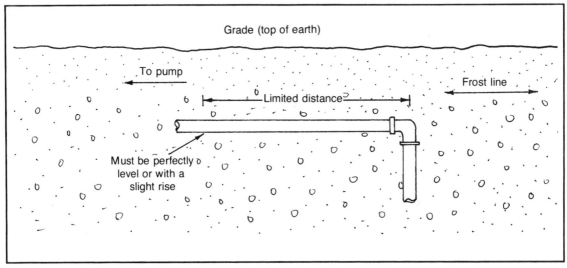

Fig. 6-7. One way a well pipe can be terminated below the surface of the earth. When a vacuum pump is used, distance to water and distance from well pipe to pump is limited.

to extract the water on a temporary basis. This can be an easily attached, hand-operated suction pump or a small motor-driven jet or similar pump with a flexible hose you can let down into the well, should the water level be below the reach of a suction pump. You don't want to stop your driving above the water and you don't want to drive the well point past the aquifer, assuming it is not very thick.

Equipment. For driving pipe, you need a cap that will go over the thread on the top end of the riser or pipe. See Figs. 6-10 and 6-11. Neither

should ever be struck directly. To do so is to mash the threads, and that makes it impossible to ever make an airtight joint.

For "hand" driving you need a sledge or a maul or a special hand driver. See Figs. 6-12A and 6-12B. The latter is a tube with an internal diameter just large enough to fit loosely over the pipe or driver end. It is usually a few feet long and half of the tube is weighted or solid.

The open end of the tube is slipped over the pipe end (with the cap in place). The tube is then

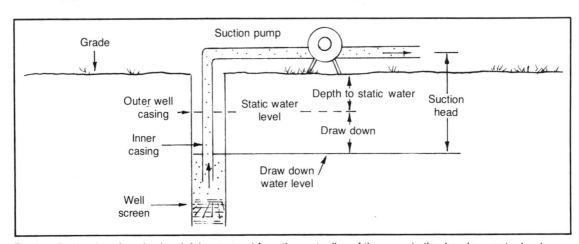

Fig. 6-8. Explanation of suction head. It is measured from the center line of the pump to the drawdown water level.

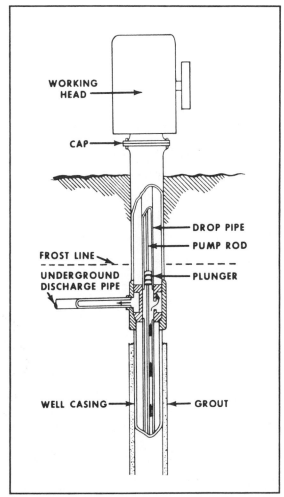

WORKING HEAD

CAP

FROST LINE

UNDERGROUND DISCHARGE PIPE

DROP PIPE

PUMP ROD

PLUNGER

WELL CASING

GROUT

Fig. 6-9. Major parts of a pitless adaptor with the pump over the well pipe.

slammed down just like a hammer. With the tube arrangement, it is impossible to miss the cap (something that is not hard to do with a sledge when you are tired). See Fig. 6-13.

For easier starting, you need a hand auger with which to make a starting hole a little larger than the outside diameter of the pipe you plan to use. Augers are discussed in Chapter 7.

Pipe Removal. For pipe removal, you need at a minimum a sturdy pipe clamp that is strong enough to take the upward strokes of the sledge and large enough to fit around the pipe size used. See Figs. 6-14 and 6-15.

In addition, you might find it helpful to have on hand two auto (or truck) jacks and a couple of lengths of 4-x-6-inch timbers about 2 feet long to serve as supports for the jacks. The timbers will keep the jack bases from sinking into the earth.

For an assist, you could also set up a sturdy tripod above the well-pipe end and hang a block and tackle or chain fall from it. Do not, I repeat, do *not* attempt to pull the well pipe with the tackle or chain fall alone. You can develop tremendous tension with even the simplest setup. Unless your tripod can withstand several tons of pressure you can easily pull the entire rig down upon yourself. Instead, tap the pipe upwards with the sledge or push with the jacks and pull with the tripod's help. Naturally, you have to move the hitch on the pipe and the position of the clamp as you ease the well pipe out of the earth.

Mechanical Assistance. In its simplest form, mechanical assistance can consist of a tripod or a derrick positioned over the top of the well pipe. See Fig. 6-16. A pulley or sheave is hung from the center of the tripod. A rope is threaded through the

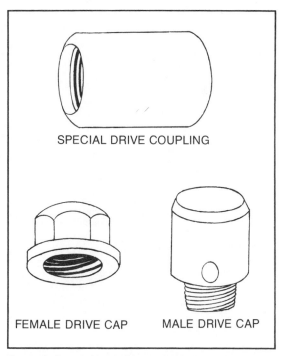

SPECIAL DRIVE COUPLING

FEMALE DRIVE CAP

MALE DRIVE CAP

Fig. 6-10. Caps and coupling used with the drive pipe.

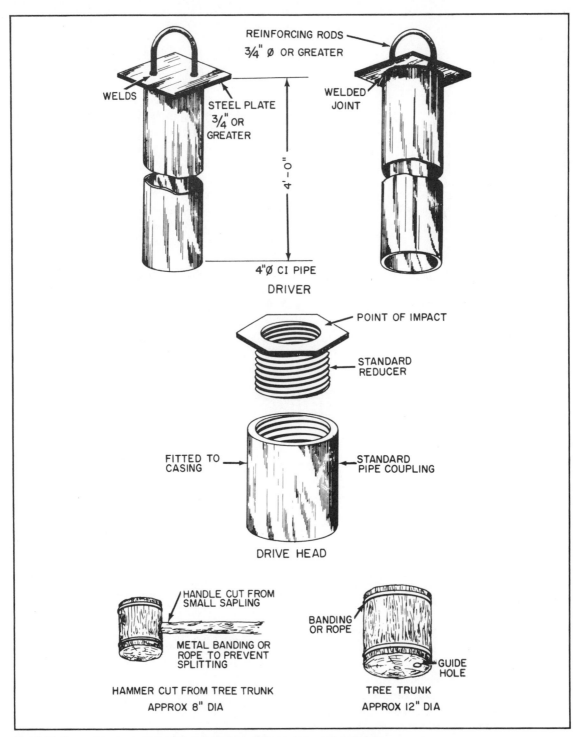

Fig. 6-11. Some drive parts that can be made from ordinary parts and materials.

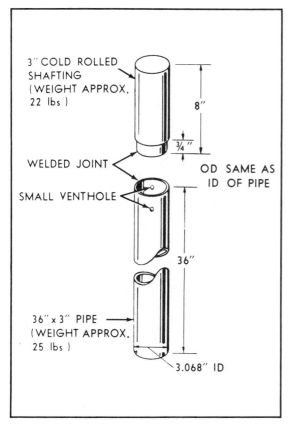

3" COLD ROLLED SHAFTING (WEIGHT APPROX. 22 lbs)

8"

¾"

WELDED JOINT

OD SAME AS ID OF PIPE

SMALL VENTHOLE

36"

36" x 3" PIPE (WEIGHT APPROX. 25 lbs)

3.068" ID

Fig. 6-12A. One driver design.

pulley. One end of the rope is fastened to a weight and the other is held in your hands. You pull the weight up, let the weight go, and gravity does the rest. Be certain to wear leather-palm gloves. The weights commonly used range from 30 to 300 pounds.

You can make a weight from a strong, steel bucket into which you pour concrete. Do not use loose stones because they will chip and fly. You can use one of the commercial weights made for the purpose; the simplest is just a weight with a flat bottom. The better types have some sort of a guidance system that is either a slot in which the pipe fits or a tube that enters the pipe top or some other arrangement. With any of the guided designs it is difficult to avoid striking the well pipe head-on. See Fig. 6-17.

The weight and pulley arrangement is much

better than the sledge. You are pulling down instead of lifting (which is much easier). You don't have to lift the hammer high to start a section and then bend low to finish it. You adjust for pipe height by simply raising or lowering the block or weight. And with the rope-and-pulley setup, two or more people can work in tandem.

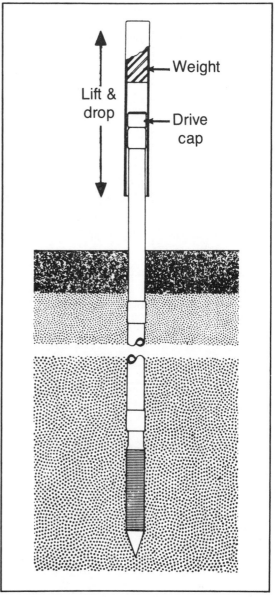

Lift & drop

Weight

Drive cap

Fig. 6-12B. How the driver is used.

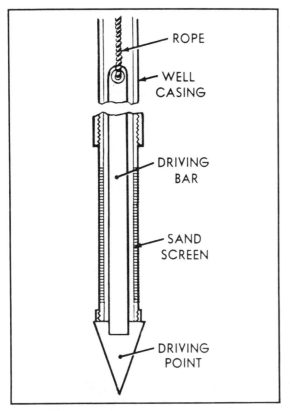

Fig. 6-13. In this well-point arrangement, the driving bar (weight) passes through the screen and strikes the top of the point directly.

Image labels: ROPE, WELL CASING, DRIVING BAR, SAND SCREEN, DRIVING POINT

Powered Assistance. A still better arrangement that removes most of the human labor involves somewhat the same setup. Instead of you or your assistant providing the motive power, you use a donkey engine to revolve a smooth drum, sometimes called a cat head, that does most of the work. This smooth, flanged drum will be 8 to 14 inches in diameter and is revolved at a rate of 3 to 6 or so rpm (revolutions per minute).

Some farm equipment comes with this geared-down drum for lifting hay into the barn and the like. A low-rpm drum can be rigged from a gas engine coupled to an old auto transmission. Set the engine at idle and geared in reverse, the wheel—without its tire—can be used as the power drum. The bare wheel is of course raised free of the ground and solidly locked into place with stakes and blocks; it cannot just stand on a jack. Note that because of the differential in the rear axle housing, the raised wheel will not rotate with any power. It will just rotate idly until it is engaged; you apply the rope. See Figs. 6-19 and 6-20.

Commercial equipment designed for this type of work has an automatic release mechanism on its drum drive. This mechanism can be set to rotate the drum just so much and then release as well as vary the speed of the drum as preferred.

Start by boring a hole into the earth with a hand

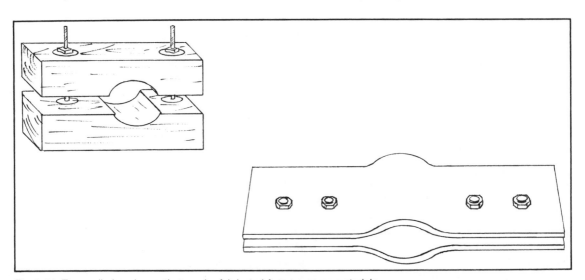

Fig. 6-14. Two well-pipe clamps that can be fabricated from common materials.

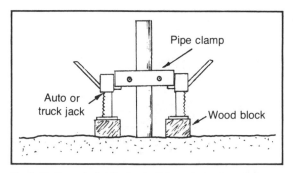

Fig. 6-15. How a pair of jacks can be used to pull a well pipe or casing.

auger. Position the auger as vertical as possible. The use of a spirit level is advised because it is easy to be fooled by out-of-vertical buildings and trees nearby. If for any reason the auger goes out of vertical, do not try to force it into the correct position. Back it out and start again. Bear in mind when you select the auger that riser size (pipe diameter) refers to internal diameter. External diameter is easily another inch and the added thickness of the couplings probably still another inch. You want a hole that will allow about ½ inch of clearance around the pipe and its couplings.

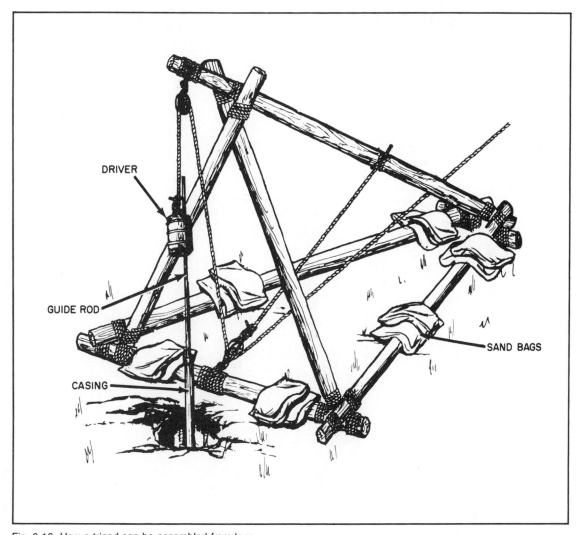

Fig. 6-16. How a tripod can be assembled from logs.

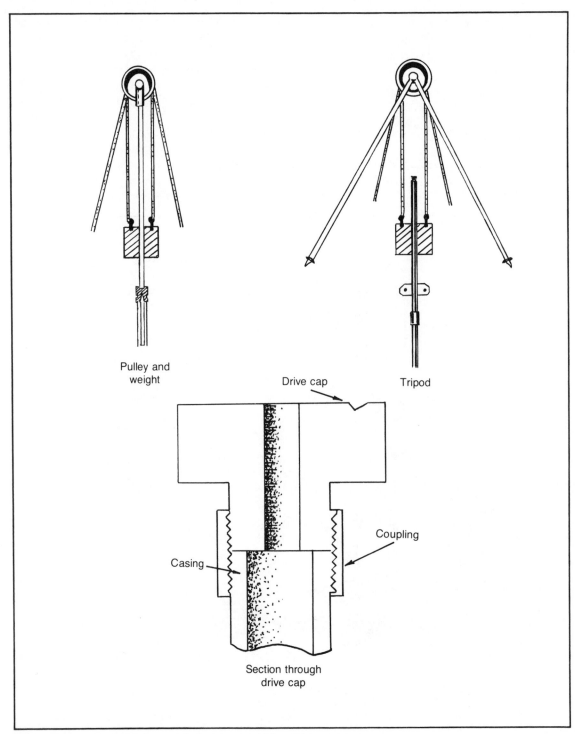

Pulley and
weight

Tripod

Drive cap

Coupling

Casing

Section through
drive cap

Fig. 6-17. Some arrangements used for keeping the driving weight centered.

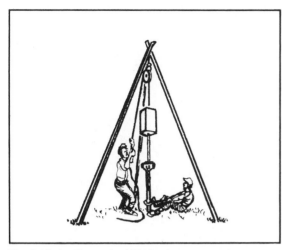

Fig. 6-18. Another makeshift tripod arrangement.

Screw the well point onto the well-pipe end. But first cover the threads liberally with pipe dope. This will make tightening the joint easier and ensure a watertight and, more important, an airtight joint. Use the two wrenches to tighten the joint. Make the joint as tight as you can without stripping the threads.

Cap the end of the first section of pipe that you are going to use. Insert the well point into the hole you have made with the auger. Check the pipe to make certain it is vertical. If it is not too much out of plumb, use dirt or wood wedges to hold the pipe vertical. If this cannot be done, bore another starting hole into the earth.

Easy Does It. With a sledge or a maul, gently tap the well point into the earth, a little at a time. At the start, you want to do no more than set a portion of the well point in the earth. Check more or less constantly in order to hold the pipe vertical. When the well point is firmly in the earth, use a Stilson wrench to make certain the well point has not come loose from its following pipe. Do *not* rotate the pipe. That might damage the well point. See Fig. 6-21.

When you are certain the initial length of pipe is firm and headed straight down, you can use more force to drive the well point. Drive the pipe as far into the earth as is convenient. Constantly check to see that the pipe has not come loose from its point. Now connect the following section of well pipe. Apply plenty of pipe dope to the pipe threads and use the two wrenches to tighten the joint. Continue driving and continue applying a little rotational force to the well pipe to make certain the sections do not unscrew.

Risers and Water Pipe. Risers are manufactured to take the pounding that is necessary to drive them deeply into the earth. Standard, galvanized water pipe cannot take such pounding. Therefore, do not plan on driving water pipe more than 15 or 20 feet. If you go this deeply and haven't reached water, you might gamble on a greater depth—if the driving is easy. If the soil resists and you find little penetration with each blow, the wiser course consists of pulling the water pipe and re-

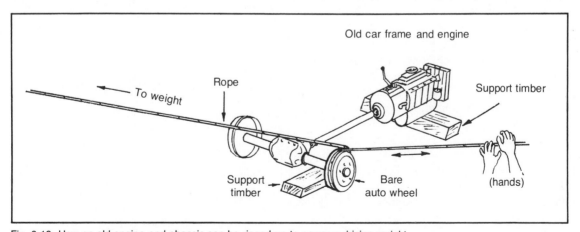

Fig. 6-19. How an old engine and chassis can be rigged up to power a driving weight.

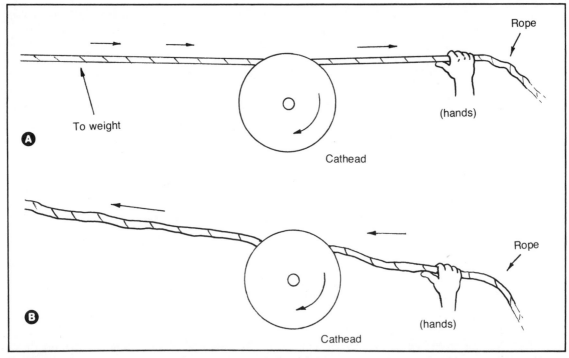

Fig. 6-20. A: When tension is applied to the rope, the rope tightens on the cathead and the rope is pulled in the direction shown. B: When the rope is released, the rope slides over the constantly revolving cathead and the rope slips, letting the attached weight drop.

placing it with the strong risers. If you try to force the galvanized water pipe, it will buckle and possibly come apart along its length. This will cost you the well point, which will now be unrecoverable.

Weights and Driving Distances. Weights or blocks commonly used for driving well points range from 50 to 300 pounds. The choice of weight depends upon the means used to lift the block, the diameter of the pipe being driven, and the nature of the soil. If you are alone, select a weight that you can handle and lift over a period of time. If you have a donkey engine providing the motive power, you can select a heavier weight. Too much weight on a small-diameter pipe or riser can mash the pipe cap. Prudence is advised.

The drop of the weight affects the pipe cap as well as the driven pipe. The farther the block drops the heavier the blow. Again, too much force can damage the well pipe and cause it to vibrate. If you drop the weight on a vibrating pipe, the blow will not be centered and might bend the pipe.

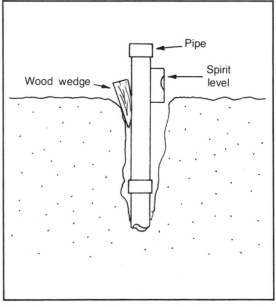

Fig. 6-21. How a spirit level and a wedge of wood can be used to hold a well pipe or casing perfectly vertical.

A light or medium weight dropped too far on a recalcitrant length of pipe will sometimes bounce up in the air. This means you have to stop and let the weight settle down (which is a loss of time). What you are seeking is a smooth rhythm. When you match the weight to the nature of the earth where the point rests and you adjust the weight height and drop time correctly, the weight will bounce upwards as you or the machine takes up the slack on the attached rope.

Centering the Block. Use the block as a plumb bob. Let it come to rest and then adjust the tripod or the derrick so that the weight can be lowered onto the pipe's cap and it will be balanced. Centering is crucial if the pipe is to be driven straight down.

Driving Rate. This refers to the distance into the earth achieved with each drop of the weight or swing of the sledge. In soft soil or sand, 2 to 3 inches per blow is about right. In "stiffer" soil the rate will be correspondingly less. In fine sand or clay, driving will be eased if you introduce water around the driven pipe.

Identifying the Subterranean Formation. A knowledge of the character of the soil your well point is penetrating is very useful. This knowledge will help you determine when you have reached the maximum practical well point depth.

This information will also help you select the best screen openings for your well point. This sounds like advice to bolt the barn door after the horse has been stolen, but it is not. Unless you have prior knowledge of the aquifer gained from neighbors or friendly commercial well drillers, you have no other way of determining the character of the aquifer except by a test drilling or judging from the sound the well pipe makes or doesn't make when it is struck, the depth it penetrates with each blow, block rebound, and whether or not the entire pipe rotates when you try tightening the pipe section with your wrench. Table 6-2 details subsoil characteristics.

If your driven well produces the flow of water you want there is no problem. If it doesn't produce enough water after developing, the trouble might be undersized well-screen openings. On the other hand, simply going to the largest openings available might fill your pump with sand. In addition to judging the subsoil character by observing the action of the block—should you decide to withdraw the well point—some of the subsoil might adhere to the well point.

This can help you select a well screen better suited to the aquifer. Before you decide that you have to "pull" the well point, try developing the well as described in Chapter 12.

Table 6-2. Subsoil Characteristics.

Type of formation	Driving conditions	Rate of descent	Sound of blow	Rebound	Resistance to rotation
Soft moist clay	Easy driving	Rapid	Dull	None	Slight but continuous.
Tough hardened clay.	Difficult driving	Slow but steady	None	Frequent rebounding.	Considerable.
Fine sand	Difficult driving	Varied	None	Frequent rebounding.	Slight.
Coarse sand	Easy driving (especially when saturated with water).	Unsteady irregular penetration for successive blows.	Dull	None	Rotation is easy and accompanied by a gritty sound.
Gravel	Easy driving	Unsteady irregular penetration for successive blows.	Dull	None	Rotation is irregular and accompanied by a gritty sound.
Boulder and rock.	Almost impossible	Little or none	Loud	Sometimes of both hammer and pipe.	Dependent on type of formation previously passed through by pipe.

COMPLETING THE DRIVEN WELL

Two general methods of terminating the well are above ground and below ground. This is your choice, but there are a number of steps to be taken before you connect your final run of pipe and call it finished.

Last Pipe or Riser. You need to plan the elevation (height or depth above the surface of the earth or below) before you attach the drive and final length of well pipe. If you desire a below-ground termination, you need to ascertain the frost line (if there is one) and dig a hole to below this depth around the pipe—plus another 6 inches of depth to give you working space. See Fig. 6-22.

If driving is fairly easy, meaning you get 1 inch or so with each blow of the block, you can safely assume that you can drive the well pipe another couple of feet without problem. In such cases you can add a temporary, additional length of well pipe to what you already have in the ground and drive the well pipe to the depth you want. Then you can remove the temporary length of water pipe or driver.

If driving is difficult, there is a chance you will reach your depth limit. The unconsolidated formation into which you are driving your well pipe will have become increasingly solid. You may just be too deep for the pipe you are using and it shows signs of buckling or you might be approaching bed-rock.

In any case, if you find the well pipe impossible to drive and it is too high for an underground connection, you have the uncomfortable choice of either pulling the well pipe back out of the hole a distance and replacing the last section with a shorter section. Or you could enlarge the hole around the well pipe sufficiently to permit you to work a pipe cutter and threader. Bear in mind that should you want to or have to cut a riser pipe short, the required threads differ from standard water pipe threads (which can be quite a problem in securing the right stock and die for the job). It is far wiser to plan ahead and secure the right length for the final section.

TESTING

You can make a preliminary test of your well before you develop and cap it. If the water level is too deep

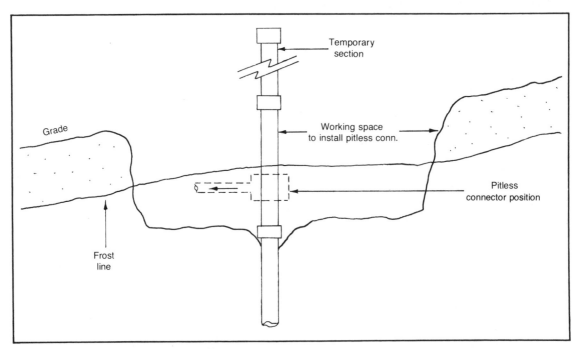

Fig. 6-22. Excavation needed to install a pitless connector.

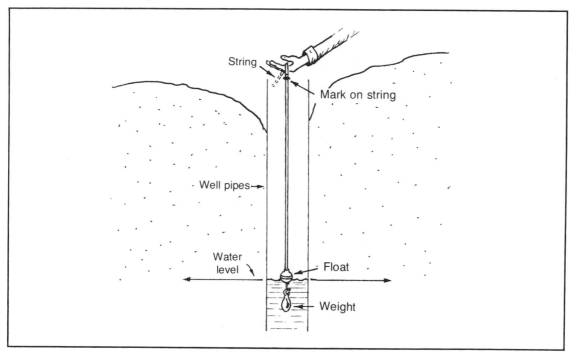

Fig. 6-23. How the water level in a well pipe can be determined.

for a suction pump, it saves time and money to know this before a pump is purchased and installed. If the flow is grossly inadequate, chances are that development will not bring the flow up to where you want or need it to be. This too is something to be considered before you do anything beyond driving the well pipe. You might need to take any or all of the steps suggested for increasing the flow of water into the well point.

Water Level. The water level can be easily ascertained with a float and weight, such as is used for fishing, attached to the end of a line. The line, float, and weight, are lowered into the well pipe. When the float reaches the water, the float carries the weight and the pull on the line is reduced. Mark the string or line at the point it leaves the well pipe. A strip of tape will work nicely. Remove the line and measure from the mark to the float (Fig. 6-23).

Flow Rate. Attach whatever kind of pump will work with the well. The pump type makes no difference except that a hand-operated pump can be tiring. Start the pump and direct the flow into a measured container such as a 55-gallon drum or a number of buckets (have a helper to empty them a distance away from the well head). Try for a continuous, steady stream of water (or as steady as you can hold it).

With a stopwatch as a guide, operate the pump for one full minute. Calculate the quantity of water drawn. Make note of the quantity. Now, without troubling to measure the quantity, operate the pump at the same speed for a steady five minutes. Remove the pump and drop the fishing line. Measure the water level. Has it dropped very much? How much? If the drop is large enough to measure, do so and note the figure.

Using your fishing line arrangement, keep tabs on the rising level of the water and note, with your watch, just how long it takes for the water in the well to rise back to its original height. Assume for the sake of discussion that the water level before pumping was exactly 20 feet below the top edge of the well pipe. Your one-minute test produced 5 gallons of water. You waited a little while and pumped for a steady 5 minutes. During this time, you removed 5 × 5 gallons of water or 25 gallons.

Quickly, you remeasured the water level within the well pipe and find that the water level has dropped 5 inches.

Because you know—or should know, having kept a record—the overall length of the well pipe to its point, you should know (by subtraction) the height of the water (before pumping) in the well pipe.

Assume that this water height is 50 inches. If you include the quantity of water that will enter the well pipe through the screen, you can pump water at the aforementioned rate of 5 gallons a minute for a total of 250 gallons before you drain the well. The figure is derived by dividing 50 inches by 5 inches (the distance the water level is lowered with the removal of 25 gallons at the rate of 5 gal/min). This produces 10, and 10 × 25 gallons results in total of 250 gallons.

Recovery Rate. Flow rate alone is not a complete measure of a well's performance. We also need to know recovery rate: the speed at which water flows into the well to replace the water that has been drawn. Assume that water entered the well at a rate of 5 gallons a minute (which it doesn't in our example). If this were the case, the well could theoretically be pumped without running dry so long as the pumping rate was held to 5 gpm (gallons per minute). But this is not the case. The water level drops 5 inches with every 25 gallons drawn at a rate of 5 gpm.

You have to measure the recovery rate to learn just how much you can draw at a steady rate. To do this, the well is given a rest, the water level is given time to reach its normal height and then the water level is pumped down a measured distance. This should be a foot or so because a shorter distance is much too difficult to measure accurately (the measurement can again be made with the line and float). The time it takes the water level to resume its original height can be measured with a stopwatch.

Assume that you pump the water level down a distance of 20 inches. The pumping rate is not important. Twenty inches of level drop represents 100 gallons; each 5 inches of drop, as previously measured, represents 25 gallons. Now time the recharge rate (the recovery rate). Assume that

exactly 40 minutes are required. Dividing 100 gallons by 40 gives you 2½.

You can pump this imaginary well at a continuous rate of 2½ gpm without running dry. You can also pump in a "burst" of 250 gallons at any rate if you then wait some 90 minutes for the well to recover.

Alternate Method. Another method that is sometimes used to measure flow and recovery rates of a small well consists of filling the well with water. The well is permitted to rest; that is to say, the water level is given time to assume its normal height in the well pipe. Then a measured quantity of water is poured into the well pipe. Any quantity can be used, but it is easier if you add sufficient water to bring its level close to the top of the pipe where you can easily see it.

Watch the water level and measure the amount of time it takes the added water to disappear. The theory behind all this is that it will take water just as long to leave the pipe through the screen as it will take water to enter. In other words, this is a test of aquifer porosity.

When the water's surface has dropped to its original level, note the time interval. Divide the time interval by the quantity of water, in gallons, that you poured into the well pipe. This will give you the rate at which you can continuously draw water from the particular well you are testing. For example, if you added 50 gallons and it required 50 minutes to drain down, your pumping rate would be 1 gpm. If the time interval was 25 minutes, the possible pumping rate would be double that.

Neither of these two methods are particularly accurate, but they are dependable and useful and will give you "ball park" numbers that will help you evaluate your well and select a pump and pumping system.

Remember that all these calculations are made on the assumption that subterrainean conditions remain stable and they do not change for better or worse. Well development will improve the flow and recovery, but there is no way to predict the improvement without actual on-site testing. The improvement might be considerable or it might be negligible. A lot depends upon the aquifer.

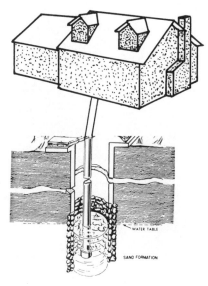

Chapter 7

Bored Wells

WATER TABLE

SAND FORMATION

BORING A WELL IS ACCOMPLISHED BY HOLDING the auger upright, forcing its point into the earth, and rotating the auger. The auger screws its way into the earth. Holes ranging in diameter from 2 to 32 inches can be made this way. Obviously, the larger-diameter augers cannot be turned by hand.

Hand-powered boring is limited to depths of 15 to 20 feet. Powered augers (Fig. 7-1) can go down to 125 feet. Boring with either hand or power equipment is best suited to homogenous formations that do not contain boulders and cobbles that have to be removed for the boring to proceed. The fewest problems are encountered when boring through clay, silt, and sand—or mixtures thereof—that will not collapse behind the auger as it descends or is removed. Put another way, boring for water wells is almost completely limited to cutting through unconsolidated formations. Standard boring equipment will not work on hard rock. Nevertheless, there are a few, very special "rigs" that will cut through consolidated formations—provided the rock is not too hard.

WHY BORE?

When you are planning to construct the well by hand, there are several factors to consider in deciding between a bored well and a driven well.

Well Diameter. It is usually much easier and faster to bore a 6-inch well than it is to drive a well point of the same diameter. The major reason for installing a 6-inch well rather than, for example, a 3-inch well is that you might want to or need to install the well pump inside the well. When the soil is mainly hard clay, it might be easier to bore than drive wells of smaller diameter.

Flow Rate. It is only natural to assume that by increasing the diameter of the well—meaning the well screen, which is the effective inlet for the well—that the flow rate will be increased proportionally. Unfortunately, it doesn't work that way. Doubling the well screen diameter increases water input by an average of only 4 percent. Well screen diameter is relatively unimportant so far as flow rate is concerned. Well yield—the quantity of water that can be continuously drawn from a well—is

Fig. 7-1. A portable, gas-powered auger. Photo courtesy Little Beaver, Inc.

much more reliant on the length of the well screen.

By rule of thumb, doubling the length of a well screen will double the yield, all things remaining equal. See Fig. 7-2. Specifically, well yield depends upon:

☐ The permeability of the aquifer.
☐ Drawdown during pumping.
☐ The diameter of the circle of influence.
☐ The length of the well screen.
☐ Rate of recharge.
☐ Well screen diameter.

The permeability of the aquifer is directly related to its composition. The coarser the aquifer the more easily water flows through it and into the well. If your aquifer is gravel, there will be more water available (all else being equal) than if the aquifer was, for example, composed mainly of fine sand or sand and clay.

Drawdown during pumping is simply another measure of the flow rate of water through the aquifer and the quantity of water present in the aquifer. If the drawdown is insignificant during pumping, the well is rapidly replenished and it can indefinitely supply the flow rate you are drawing.

The circle of influence is a measure of the rate at which water enters the aquifer surrounding the well. If you pump your imaginary well and the water table drops for a few thousand feet in every direction—a circle of influence around the well— you know that water is not entering the area around your well as fast as you are pumping. The other hand, if the radius of the circle of influence is comparatively small, you know that the aquifer is well supplied with replenishing water. See Fig. 7-3.

If you were to drill and redrill a number of wells of increasing diameter in the very same

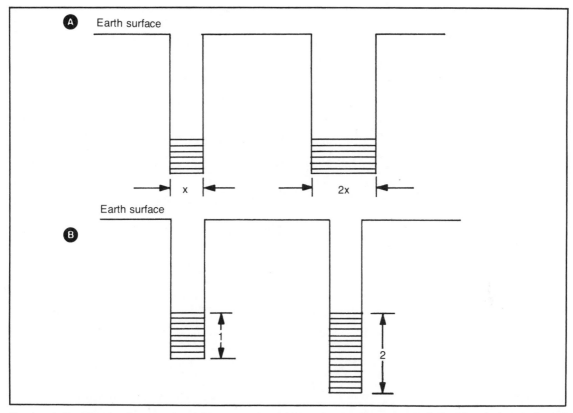

Fig. 7-2. A: Doubling the diameter of a well increases the yield by only a small percentage. B: Doubling the length of a well screen doubles yield. See Fig. 7-3.

spot—going from a 4-inch diameter well screen to a 12-inch well screen, for example—would increase water yield only 12 percent. The relative costs of a 12-inch well as compared to a 4-inch well would easily be on the order of 5 to 1.

Packing. Packing is the term used to denote crushed stone, sand, or gravel positioned between the well screen and the surrounding earth. See Fig. 7-4. Under certain conditions, the packing can effectively increase the effective inlet area of the screen and so increase well yield. It all depends upon the composition of the aquifer. It is most often used in fine sand and silt. Packing is rarely used in coarse gravel and coarse sand. Many well experts believe that packing is not worth the time and effort, and that extensive development work can produce as much yield from a well as packing it.

It is very difficult to pack a driven well. Packing a bored (or drilled or dug well) is not especially easy either, but it is done on a routine basis with little problem. Packing techniques are discussed in Chapter 12.

HAND-TOOL BORING

If the top layer of earth is soft, it is best to hand dig through this layer. If you attempt to start the auger (Fig. 7-5) in topsoil, it will tilt and fall over should you let it go.

Upon reaching the firmer subsoil, the point or working end of the auger is forced into the earth. The auger is positioned vertically and all you need do is turn its handle. When its blades are filled with soil or as much soil as you can comfortably lift, the auger is lifted from the earth, the soil is dumped,

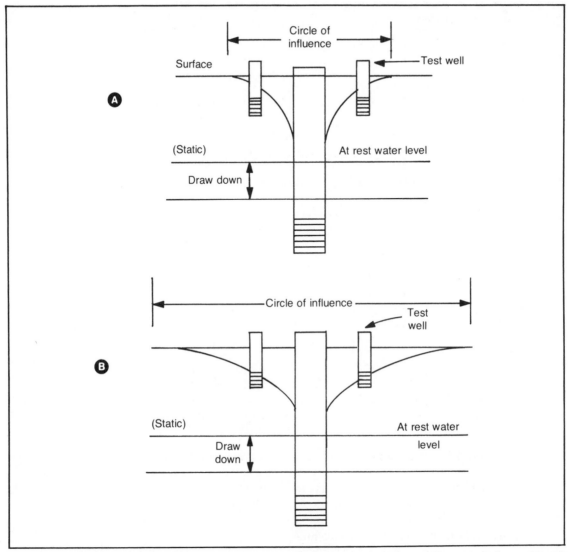

Fig. 7-3. A: In highly permeable aquifers, drawdown does not drain the water from a large area. The area of influence is small. B: In less permeable aquifers, drawdown is accompanied by a large circle of influence.

and the auger is replaced.

Depending upon the tenacity of the soil, your arm strength and the depth of the auger, there will come a time when you will no longer be able to turn it by its handle. When this happens, you can try backing the auger out, turning it as you go to enlarge the hole, and loosening the tool in the earth. When that trick no longer works, you can extend the handle with a length of pipe. The thing to be cauti-

ous about here is that if you use a very long length of pipe you can develop sufficient torque to either break the handle or destroy the shaft.

Extensions. When you add one or more extensions to the earth auger, its length becomes difficult to handle standing on the ground. The first remedy is to provide yourself with some sort of sturdy platform upon which you can stand. When you have reached a depth in the earth where standing on a platform doesn't help you pull the auger each time it is filled with earth, you need to set up some sort of tripod or derrick, along with block and tackle, to help with the extended auger's weight.

Cobbles. Small stones will fit into the tines on the auger and can be lifted out with the dirt. Larger stones—cobbles or cobblers, as they are called—have to be removed one at a time. This can be accomplished with the aid of a spiral or ram's horn auger. The dirt auger is removed and the spiral auger is lowered and carefully turned until the obstruction is caught in the spiral and it can be lifted out of the bore. See Fig. 7-6.

Should you encounter a larger boulder, head-on, there is nothing for you to do but withdraw gracefully and try again nearby. There is no way you can remove any stone larger than the bore hole using hand tools.

WORKING IN COLLAPSING SOIL

Making a bore hole in soil that will not close upon the opening as you remove the auger is merely a matter of work. When the soil collapses into the bore behind the auger or as soon as you remove it to clear the hole, you have two courses of action open. You can case the hole as you proceed or you can abandon this technique and sink your well by jetting it. Jetting is discussed in Chapter 8.

Casing the Well. In firm soil, you case the bore hole after you have drilled it to the desired depth. With this technique, the well is cased as you bore it. As you can imagine, casing while boring is much more difficult.

The procedure is as follows. Select a steel casing with an internal diameter at least one-half inch larger than the external diameter of the auger you plan to use. You need this clearance to prevent the auger from "freezing" in the casing.

Bore a hole a foot or so into the earth. Remove the auger and position the casing vertically in the bored hole. See Fig. 7-7. Some sort of cap is placed atop the casing pipe. Now, with a sledgehammer, the casing is driven into the earth as far as it will go without danger of damage. The cap can be a pipe

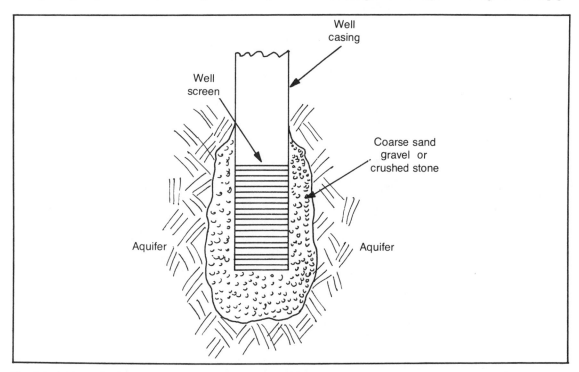

Fig. 7-4. An example of packing. A layer of crushed stone around the well screen can increase the yield of a well.

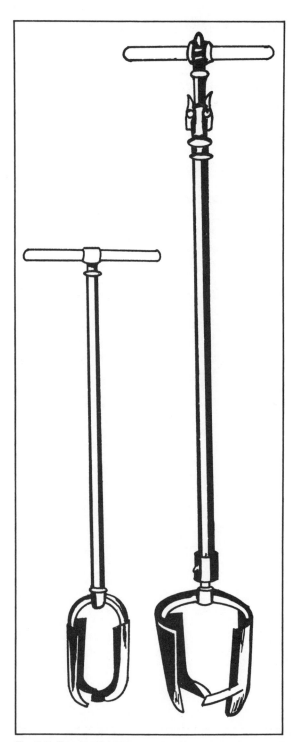

Fig. 7-5. Two-hand augers.

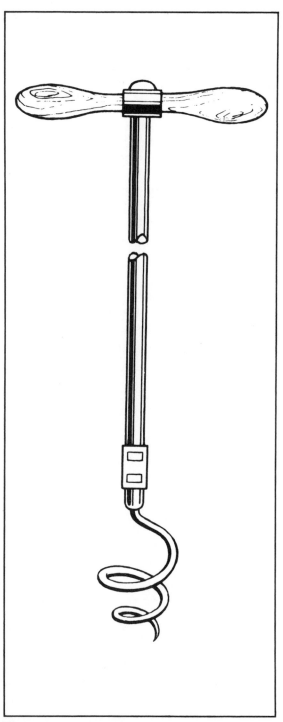

Fig. 7-6. A spiral auger used to remove stones from the bore hole.

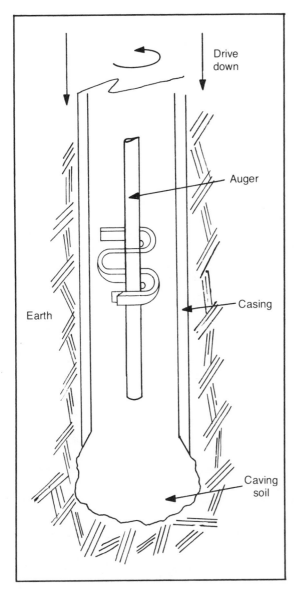

Fig. 7-7. A well casing used to hold caving soil in place.

spirit level is applied to its side to make certain it is perfectly vertical. If it is not, straighten it. The auger is then slipped down through the center of the casing and used to remove some more soil. The auger is removed and the casing is driven into the earth again.

All this is possible because the soil in which you are boring tends to cave in. Although the hole the auger is making is smaller than the well casing you are driving, the hole's diameter increases somewhat when the auger is removed. This is a tedious, time-consuming job.

You can turn the casing as you drive it with the aid of a pair of pipe tongs (Fig. 7-8). Obviously, two well drillers are required to drive the casing and turn it more or less simultaneously.

Well Completion. The completion of a well involves the "setting" of a well screen at the lower

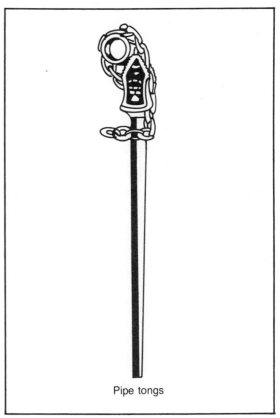

Fig. 7-8. Pipe tongs are a kind of pipe wrench.

fitting or a block of 2 × 4. It helps to cut saw teeth into the bottom edge of the pipe before you begin, of course, and sometimes a little water poured around the casing helps lubricate it and makes it easier to drive. Too much water turns the soil to mud and the auger cannot lift mud.

When the casing has been driven as far as it will go without mashing its top with the sledge, a end of the casing. This is covered in Chapters 11 and 12, as is the subject of well packing.

Depth. In theory, you can bore through to China if you have a sufficient number of auger extentions. In practice 20 feet is about maximum. Beyond this depth, the torque needed to turn the auger reaches a point where the auger will be twisted or its handle will break.

NONCOLLAPSING OR NONCAVING SOIL

Unless the soil is firm clay or a mixture of clay and sand and or silt right to the very surface of the earth, it is best to dig a pilot hole by hand to get rid of the soft topsoil. Then it is just a matter of turning and lifting as previously described.

Casing a Bore Hole In Noncaving Soil. When boring in soil that caves in behind the auger, select an auger smaller than the internal diameter of the casing pipe you plan to use. Working in noncaving soil, select an auger that will bore a hole *larger* in diameter than the outside diameter of the casing pipe by at least 1 inch (and even more if you plan to pack the well screen).

In either case, the slightly oversize hole is bored as straight and as vertical as possible. If the hole is crooked, you will not be able to position the pipe inside it.

After the hole has been bored, the casing is lowered until it rests on the bottom of the hole. As many casing sections as required are used. They can be joined either by electric arc welding or by screw-thread couplings. Pipe suited to either methods of joining must be used. The last length is precut, if necessary, to terminate the well casing at the desired height.

The hole diameter for a well that is to be packed should be twice the desired packing thickness plus the outside diameter (OD) of the casing. For example, if you wanted to surround a 6-inch OD with a 2-inch layer of packing, you would bore a hole with a diameter of 10 inches. Bear in mind that pipe is usually specified by its inside diameter. The OD of a pipe depends upon wall thickness and this varies from one grade of pipe to another. To be certain of the OD of the casing pipe you plan to use, measure its OD.

POWER AUGERS

There are two types of power augers: The continuous auger is shown in Fig. 7-1. This unit is made by the Little Beaver Company of Texas. It can be disassembled and carried in the back of a station wagon or a light truck. In some areas of the country, it is possible to rent a machine like this. It can bore to a depth of 18 feet with an auger 4 inches in diameter. Its auger is in the form of a continuous spiral. As the auger turns, it digs itself into the earth and deposits the soil alongside.

The other type is classified as a bucket auger. It consists of a bucket with an open bottom to which are fastened teeth that act as an auger. The bucket is attached to a rigid stem that passes through a shaped hole in a power-driven rotating table. The rigid stem with a square or hexagonal cross section is called the kelly. When the table rotates the kelly rotates with it. At the same time, the kelly and its attached bucket are free to be lowered and raised as required.

Capabilities. Working in an unconsolidated formation, an auger bucket can bore a hole at an average rate of 30 to 40 feet per hour and, in some instances, as much as 60 feet in one hour. When conditions permit, commercial drillers often excavate five or six wells per day. Bore hole depth is limited to the length of the kelly and the number of extensions that can be added on. The "Super George" bucket auger made by the Gus Pech Company, Le Mars, Iowa goes down to 125 feet. Maximum diameter of the bored hole is usually 30 inches. This permits the bore hole to be lined with 24-inch ID concrete pipe.

The bucket auger works best in unconsolidated formations. Such formations will not cave in behind the bucket. When slight or mild caving is encountered, many rotary, bucket-drill operators fill the bore hole with water, depending upon the hydrostatic pressure to keep the walls from collapsing.

This might seem strange or erroneous in the light of the previous statement concerning hand augers (too much water will turn the soil to mud and the hand auger will not operate in mud). The difference is partially due to the speed at which the bucket auger is rotated and partially due to the shape of the bucket. In many ways, the bucket acts like a scoop. When it has scooped up the mud and continues to turn rapidly enough, the mud remains in the bucket. This technique works best in sand.

Casing the Rotary-Bucket Bore Hole. Casing is necessary where the soil is so soft that it will not stop caving even when the bore hole is filled with water. The usual rotary bucket is replaced with a bucket that has "expanding" teeth on its bottom edge. When this type of bucket is used to bore a hole, it makes a hole several inches larger than the diameter of the bucket itself.

A hole is bored as deeply as it is possible to go without the hole walls caving in. Then a length of casing is inserted into the bore hole and positioned vertically. The bucket auger, attached to its kelly, is lowered through the casing until it reaches the soil. The bucket is revolved.

When sufficient soil has been removed below the casing edge, the bucket's teeth are freed from the confines of the casing and they expand. Now the bucket cuts a hole larger in diameter than the casing. The casing will move down of its own weight or can be easily driven down.

The use of a well casing for boring increases the cost of well material and labor. If you prefer, the casing can be left in place and used as the well casing or pipe. Or a second casing (call it well pipe) can be inserted in the in-place casing used as the means of drawing water.

Procedure. The site selected must be such that it can be reached by the truck bearing the boring rig and it must be sufficiently level or capable of being leveled so that the rig can be leveled.

The rig is driven to the site, and its back end is positioned over the location of the desired bore hole. The hydraulic outriggers are energized and the truck body and its machinery are lifted off the ground and made perfectly horizontal. The mast or derrick is raised and locked in place (which makes it perfectly vertical). This too is accomplished by hydraulically powered pistons and cylinders.

The telescoping kelly is lifted into position and one end is lowered through the hole in the rotating platform. This is part of a large ring gear driven by a pinion gear in turn attached to the *draw works*. The latter is the term used to describe the engine and attachments that drive the rotary table and which can be used to "draw" (lift) pipe and whatever as required.

The bucket is now attached to the bottom of the kelly. The kelly is lowered until the bucket rests on the ground, and then the table is spun and the bucket digs into the earth. When the bucket is full, as witnessed by the driller, the kelly is retracted and lifted free of the turn table and brought to one side of the machine where its soil is simply dumped for later removal.

Soil Sampling. One tremendous advantage the bucket boring and even the hand-boring method of constructing a well has over other techniques is that every time you bring up a bucket of soil, you are sampling the formation. Thus you know almost immediately when you strike the water-bearing formation and when you have unwittingly gone past it. The latter condition presents no serious problem because the lower portion of the bore hole can be sealed with several feet of gravel, crushed stone, or even concrete.

Casing an Open-Bore Hole. The hole is bored as previously described. The casing is lowered into place with the aid of a *trip line* (a line with an automatic release mechanism). This is done one section of casing at a time. The first section is suitably pierced to permit the entrance of water. When the well is to consist of but a single casing, the bore hole is made oversized by several inches, and the space between the casing and the aquifer is packed with coarse sand, crushed stone, or gravel.

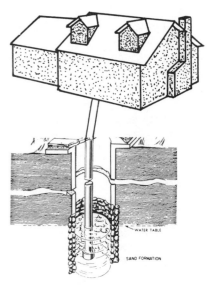

Chapter 8

Jetted Wells

WATER TABLE

SAND FORMATION

J ETTING IS A METHOD OF CONSTRUCTING A well using the flow of water to bore the hole in the earth. If you were to connect a garden hose to a source of water, turn the water on, and place the open end of the hose against the earth, the moving "jet" of water would produce a hole in the earth. If you continued to press the end of the hose into the hole, the force of the water would loosen the soil and sand at the bottom of the hole and the movement of the water would carry it up and out of the hole. As soil and sand are removed, you could continue lowering the end of the hose more deeply into the hole. This would make the hole deeper and deeper. See Fig. 8-1.

This is the basic principle behind jetting a well. In actual practice, the garden hose is replaced by a stiff pipe. The flow of water can be aided by a flow of compressed air. The pipe can be rotated and it can also be driven. The jet can be used to form a hole for itself or it can be used to provide a hole for a well casing.

Jets of water and air are also used in conjunc-

tion with rotary and percussion drilling. This chapter describes well-digging techniques that depend mainly on the flow of water and sometimes air for excavation.

MAJOR PARAMETERS

There are no specifics outlining the quantity of water needed, the required pressure, or the rate of penetration possible with jetting. The larger the desired bore hole, the more compact the soil structure, the deeper and faster you want to go, the greater the volume of water required and the applied pressure.

Water Pressure. Generally, water at 50 psi (pounds per square inch) is adequate for most work. But this is no reason not to try with whatever water pressure is available (even water taken from a residential source). The only result of a lack of pressure at the lower end of the jet is to fail to lift the soil and sand out of the jetted hole. Low pressure will also limit the depth to which your jet will work effectively. But no harm will be done.

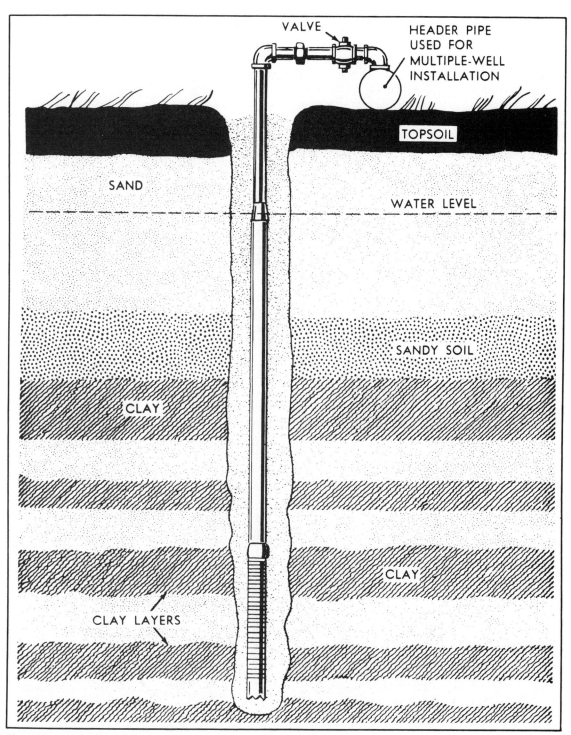

Fig. 8-1. A jetted well in place.

If you are going to rent a pump to provide the required water pressure, you must first determine the quality of the water you are going to use. If it is to be fresh, clean water, you can use any type of pump that is large enough. If you are going to use recirculated water, which is what is normally done when there isn't an unlimited supply of clean water at hand, you must use a centrifugal pump. An ordinary piston pump would soon be ruined by the mud and sand in the pumped water.

Water Volume Required. Again, speaking generally, a volume of some 50 to 100 gpm (gallons per minute) is sufficient for most work. Again, the practical approach is to make a test run with what you have.

Pump and Driver. For best results, the electric motor or gasoline engine driving the pump should be matched to the pump. If the pump is too large, the driver will be overloaded. If the pump is too small, it will be rotated beyond its designed range—cutting its efficiency. In most instances, when you purchase or rent a pump and drive set the units are properly matched. Note that unless the source of water supply furnished the pump is higher than the centerline of the pump, the pump has to be primed. This means that there has to be water in the pump before it will draw (pull water in and drive water out). See Fig. 8-2.

Bore Hole Diameter. The diameter of the hole the jet stream makes is always larger than the outside diameter (OD) of the jetting pipe. Typically, a 1-inch pipe will produce a 2-inch hole. A 1½-inch pipe will produce a 3-inch hole and a 2-inch pipe will produce a hole 4 to 6 inches in diameter.

Jetting Speed. Some drillers working with simple jetting equipment, but with sufficient water under sufficient pressure, report that they can jet a small-bore hole to a depth of a dozen feet in a few minutes. In heavy formations, jetting alone will not produce penetration (or perhaps at a very slow rate). Under such conditions, penetration is only achieved at a reasonable rate when the jet pipe is also driven.

Jetting Depth. The jetting system of well-hole excavation is most frequently used for shallow wells that are on the order of 25 feet or so in depth. Soil structure permitting, jetting is even a faster way of sinking a well than driving it. Maximum depth possible with jetting alone is limited. The exact figure depends upon the nature of the formation. With moderate driving, well casings have been sunk to several hundred feet.

Bear in mind that the deeper you go the greater the water pressure required not only to scour the hole but to bring the soil and sand to the surface. Remember that the weight of water alone amounts to 0.43 pounds per square inch per vertical foot. Put another way, to lift water 1 foot you need a pressure of 0.43 psi. Thus you need 4.33 psi to lift water 10 feet or 44 psi to lift water 100 feet.

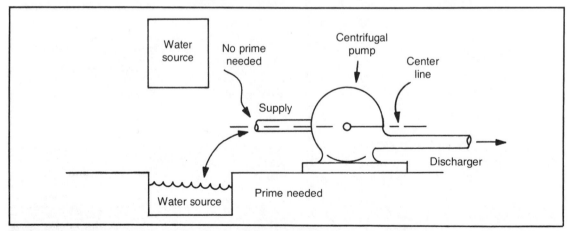

Fig. 8-2. The water source should be above the centerline of a centrifugal pump, and it has to be primed before it starts pumping.

This pressure is merely that necessary to return the water in the bore hole from that depth. It doesn't include friction loss or the force needed to free the sand and soil. Sand usually requires less pressure but more water. Clay usually requires less water but more pressure. Sometimes as much as 150 psi is needed to move clay and lift gravel.

Note that if the single centrifugal pump you have will not do the job, you can connect two in series. For satisfactory results, they must be similar in capacity. If they are not, have the larger unit feed the smaller.

The volume of water needed can be roughly computed on the basis of bore-hole depth and its actual diameter. Use the standard formula for computing the volume of a cylinder. Double the figure to account for leakage into the formation and waste.

MAKING YOUR OWN JETTING EQUIPMENT

A jetting tool can be quickly constructed from a length of ordinary water pipe with an inside diameter of 1 inch or more. One end is almost flattened to make a slotlike opening about ¼-inch wide. The exterior edges of this end of the pipe is then ground down to a coarse point with the aid of a power grindstone, producing a smooth chisel point. The

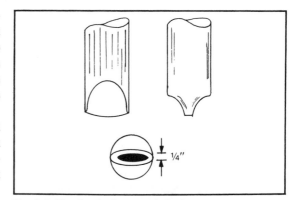

Fig. 8-3. The tip of a homemade jet lance.

other end of the pipe is naturally threaded. To this end, an elbow (right-angle fitting) is screwed on, and in turn a nipple (short length of threaded pipe) about 1 foot in length is screwed fast. The water hose is then fastened to the end of the nipple. That is all there is to making a simple jetting tool with ordinary pipe and hose. See Figs. 8-3 and 8-4.

To penetrate more deeply into the earth than the single section of pipe will permit, you have to unscrew the elbow and extend the jet, (often called the *lance* or *jet lance*). This can be repeated as often as necessary to go as deeply as you want or water pressure and available water will permit.

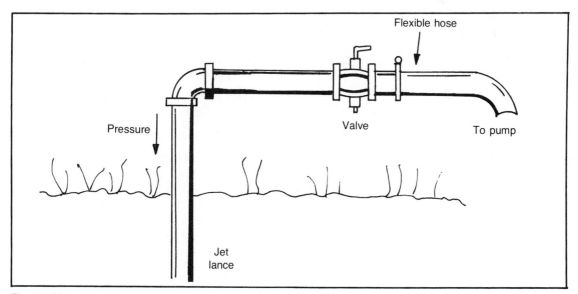

Fig. 8-4. How the jet lance is used.

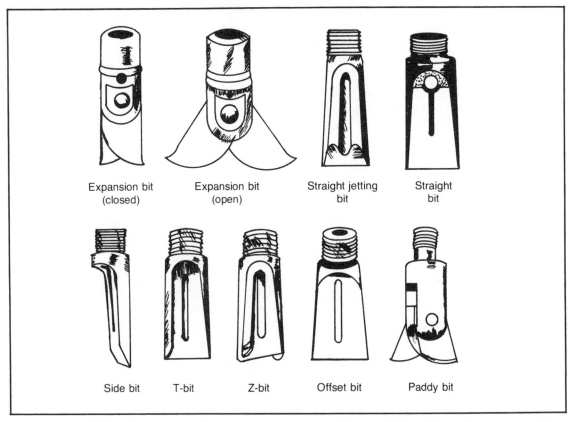

Expansion bit (closed)   Expansion bit (open)   Straight jetting bit   Straight bit

Side bit   T-bit   Z-bit   Offset bit   Paddy bit

Fig. 8-5. Commercial jet-lance bits. The expansion and the paddy bits are designed to be slipped through a casing or well pipe.

An alternative to the homemade jet lance consists of the same water pipe to which a commercially manufactured bit (tip) has been screwed on. The first point to bear in mind when selecting a commercial bit is to make certain that it is made for the pipe diameter you have or plan to use. The second is to make certain that the screw threads on the bit you want to purchase will match the common pipe thread to be found on galvanized water pipe. If the threads do not match, the bit will not stay in place. If you have to secure pipe with threads suited to your bit, you will have to do the same for the nipple and the valve (unless you can secure an adapter that will permit proper joining of the two different threads).

Figure 8-5 shows several types of jet bits. The expansion bit and the paddy bit are designed to be slipped through a pipe casing, expand and jet a hole considerably larger than the casing, and enable the casing to be more easily jetted in place.

Another jetting tool that can be home constructed is shown in Fig. 8-6. This device is used when and where you want to jet and install the well casing in one operation.

USING THE JET LANCE

The initial problem when jetting a well with what might be called hand equipment is making certain the jet tool or lance is perfectly vertical during its first few feet of penetration. This is not at all difficult if there are two workers on the job. Working alone, it often helps to start by digging a hole in the earth a foot or so in depth, and by constructing the simple positioning guide as shown in Fig. 8-7.

Jetting. The lance is positioned vertically, its tip rests on the soil. Water pressure is turned par-

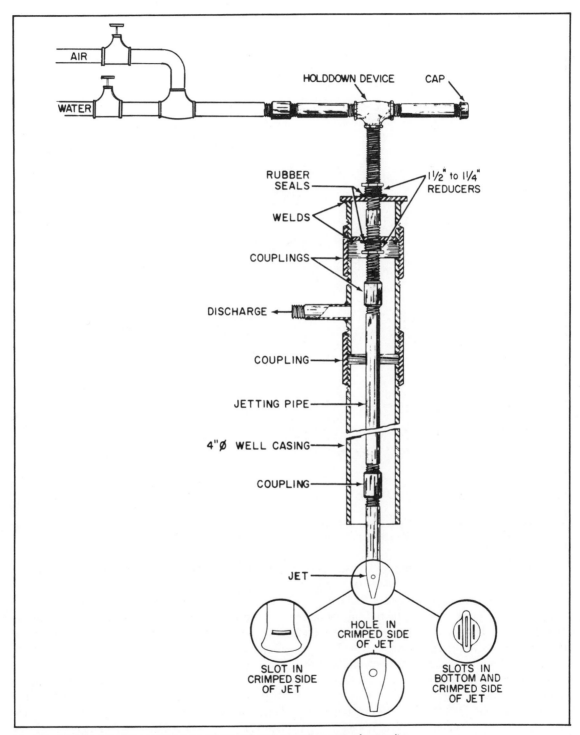

Fig. 8-6. A jetting arrangement that uses compressed air as well as water for results.

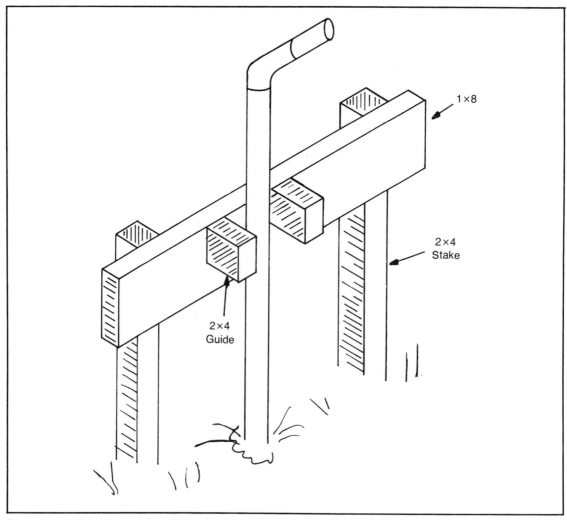

Fig. 8-7. A simple guide that will hold a jet lance in a vertical position while it is being used.

tially on. The lance is turned on its axis from right to left and back again to direct the water stream in an even flow. When the lance tip has entered the earth to the depth of 6 inches or so, the turning action is complemented by a gentle lifting and lowering of the lance; again to produce a more even flow.

As the lance continues to bore its way into the earth, the previously described actions are continued. As you penetrate more deeply into the earth, there will come a point where the water flowing out of the annulus will have little color and little debris. This indicates that either the water

pressure is too low or the formation is too hard for the available pressure. Try increasing the pressure. If this does not materially increase the rate of penetration, try driving the jetting lance. Do this by lifting the lance and attached pipe a foot or so and dropping it. Then turn it and repeat the operation. See Fig. 8-8.

When you have reached a point where vigorous driving and full water pressure fails to advance the lance into the formation, you have either reached bedrock or are at the limit of your equipment. If the water outflow suddenly stops or dwindles to almost

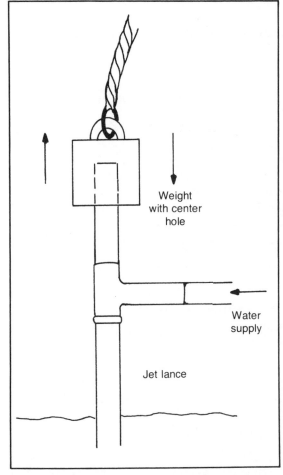

Fig. 8-8. A weight can be used to drive a jet lance while the flow of water washes soil from beneath the tip.

nothing and the pump is operating normally (even running a bit faster than usual), the orifice at the end of the lance is plugged up. This is of course due to the poundings you have given it.

Cut the pump and remove the lance, but be careful to keep the bore hole filled with water. Then clean out the orifice. If clogging becomes a nuisance, drill a number of ⅛-inch holes in the lower end of the lance. Position these holes at the outside of the pipe so that the jets travel horizontally—at right angles to the axis of the pipe.

Observations. Keep close tabs on the debris brought to the surface by the stream of water. The nature of the debris will help you determine when

you have reached water-bearing sands. When you have penetrated to a sufficient depth to either center your well screen in the thickness of the aquifer (the water-bearing sand) or sufficiently deep to position the well screen several feet below the top edge of the aquifer, you are ready to complete the jetted well.

COMPLETING THE LANCE-JETTED WELL

At this point, you have jetted a length of pipe the preferred depth into the formation. So far as you know, you need penetrate no farther. The first thing to do now is turn down, but not off, the water supply. You want a slow, gentle stream of water entering and leaving the bore hole.

Next, prepare the well screen and the well pipe. The same type of well point that is described in Chapter 6 is screwed fast to a length of pipe. The joint is first given a coat of pipe dope and then made up tight. The additional lengths of pipe that will be required are placed at hand and made ready.

The lance is removed and the well screen (well point) and attached pipe are lowered into the bore hole. Then the required following lengths of pipe are screwed fast as required. If you have a long string of pipe to "make up," it is advisable to arrange for some sort of clamp to hold the weight of the string of pipe while you are attaching more pipe.

A 10-foot length of 1-inch ID pipe weighs 17 pounds. The same length of 2-inch pipe weighs 37 pounds. If you are down to a modest depth of 40 feet and running a 1-inch line, you are dealing with 68 pounds when you couple the last length of pipe in place. With 2-inch pipe, you would be trying to hold back 136 pounds.

Working in Caving Soil. Again, this section deals with the use of the simple jet lance. The first indication that the soil is caving in on the lance is that the sides of the bore hole collapse to some degree even while the jet of water is flowing and the lance is in the bore hole. Another indication might be the closing of the bore hole under the end of the lance. The lance is lowered and rotated and possibly driven (by the force of lowering it) into the formation. The lance makes considerable penetration. But when the lance is lifted and dropped the following time, there seems to be no penetration. The lance is at the same depth or almost the same depth. This is due to the walls of the bore hole caving in and filling the bottom of the hole.

One solution is to keep the bore hole filled with water at all times. During the period the jet lance is removed, the hose should be directed to keep the bore hole filled with water. In many instances, the water level in the bore hole will drop rapidly once the supply of water is shut off.

Another answer to the problem of caving consists of holding the well-point screen and its attached well pipe against the side of the jet lance. See Fig. 8-9. Both lengths of pipe are then moved up and down together. In this way, the jet stream opens the path for the well point. Even though the

bore hole walls will collapse, they collapse upon the well point. After both pipes have been jetted to the preferred depth, the jet lance is removed. Note that removal will be made easier by continuing to let the water flow as you raise the jet lance. To keep unwanted soil from coming up out of the bore hole along the lance, water pressure is reduced as the lance is raised.

Self-Jetting Well Points. In most respects, the self-jetting well point is exactly like the well points shown in Chapter 6. They consist of a screen with provisions for attachment to a riser pipe at one end and a reinforced point at the other end. Unlike the driven well points, the bottom end of the self-jetting well points have one or more openings leading to a one-way valve that will permit water to flow

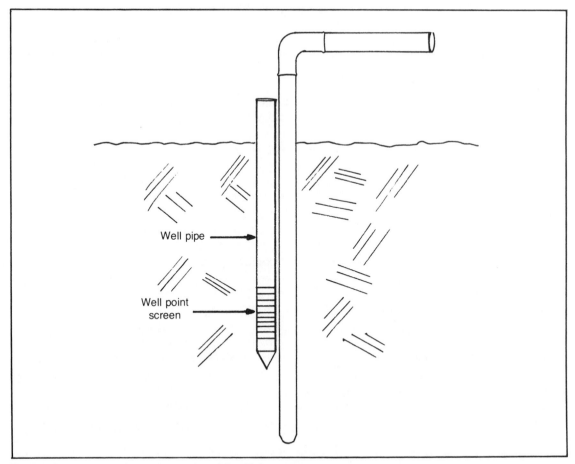

Well pipe

Well point screen

Fig. 8-9. A jet lance can be used to wash a well point into place.

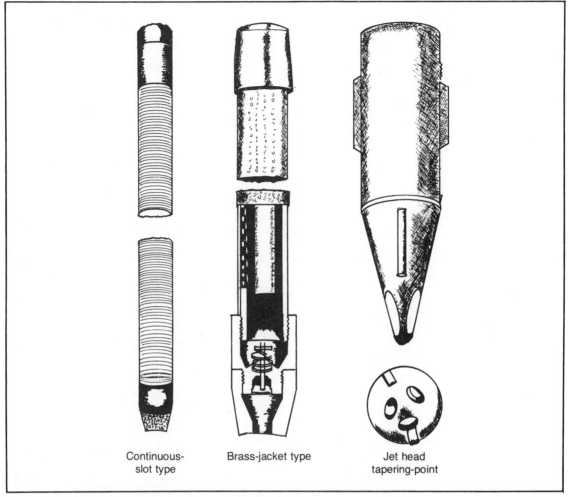

Continuous-slot type   Brass-jacket type   Jet head tapering-point

Fig. 8-10. A few self-jetting well points. The valve at the tip permits water under pressure to pass through and wash a path for the well point. When water pressure is removed, the valve closes.

down and out of the holes but will not let water return. The valve is spring loaded. See Fig. 8-10.

As you can see, the self-jetting well point has a lot of advantages, but it does have one serious drawback that makes it far from first choice when selecting fast and inexpensive well-construction equipment. The self-jetting well point requires a second pipe that is slipped down the length of the outer pipe. With but a single pipe, water pumped down to the self-jetting well point would flow out the screen. See Fig. 8-11. Instead of a high-speed flow of water through a restricted orifice, there would simply be a large volume of water slushing

through the screen. It would not cut into the formation.

The self-jetting well screen, attached to its pipe, with a second inner pipe attached to the water supply is worked just as previously suggested for the jet lance alone. Both pipes, the inner and outer, can be raised and lowered as a unit, or the inner pipe alone can be raised and permitted to drop to provide the driving force. The outer pipe and its attached screen can be rotated very gently to and fro to preclude stones from catching in the screen and ripping it open.

Driving the Jet Lance. To speed the rate of

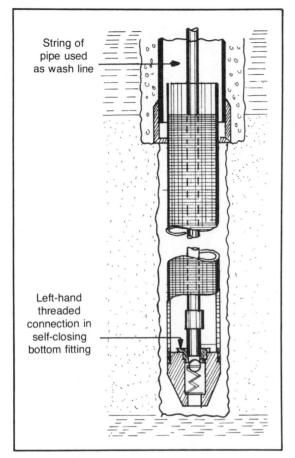

String of pipe used as wash line

Left-hand threaded connection in self-closing bottom fitting

Fig. 8-11. Inside one type of self-jetting well point. After the jet has washed its way down to the preferred depth, the inner pipe is unscrewed.

penetration and possibly to pierce dense clay formations, you can opt to apply more downward force to the lance than is possible by merely using the weight of the pipe alone. When you have a long string of pipe in the ground, it will be easier to drive the lance with a falling weight than to lift the entire string and drop it.

There are any number of ways this can be accomplished. One way is to insert a T fitting and a nipple ahead of the top end of the lance. A cap of some sort such as suggested for the top of a driven well point is positioned atop the lance. The water hose is now connected to the T fitting extension. This leaves you free to "pound" away at the top of the pipe. Again, any of the driving methods and gear

described in Chapter 6 can be used. The nipple on the T fitting can be used as a handle to turn the lance to and fro.

WASHING IN AND DRIVING A CASING

A casing is used in the construction of a well when the formation will not remain open while you jet it or when you remove the jet and attempt to insert the well pipe. In other words, there is little choice but to install a casing when the soil caves. Casings are used when a well-pipe diameter is required that is not easily jetted into place. Casings are used when jetting alone will not penetrate. A casing can be driven safely with much more force than can a well point.

Casing ID. A casing diameter is selected with the preceding restrictions in mind and also in relation to the outside diameter of the well screen to be used. The well screen should be of the telescoping type; it is telescoped (slipped inside) the casing and positioned beyond the lower edge of the casing. Following, the upper edge of the screen is expanded. This is discussed more fully in Chapter 11.

Washing in a Casing. Washing in a casing can be accomplished by digging a small hole in the earth to facilitate starting or even boring down a few

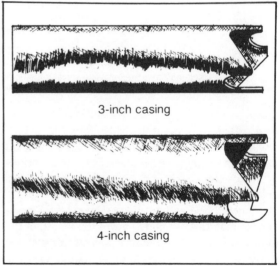

3-inch casing

4-inch casing

Fig. 8-12. Cutting edges formed on the end of a casing. These teeth help the casing cut downward when the casing is rotated.

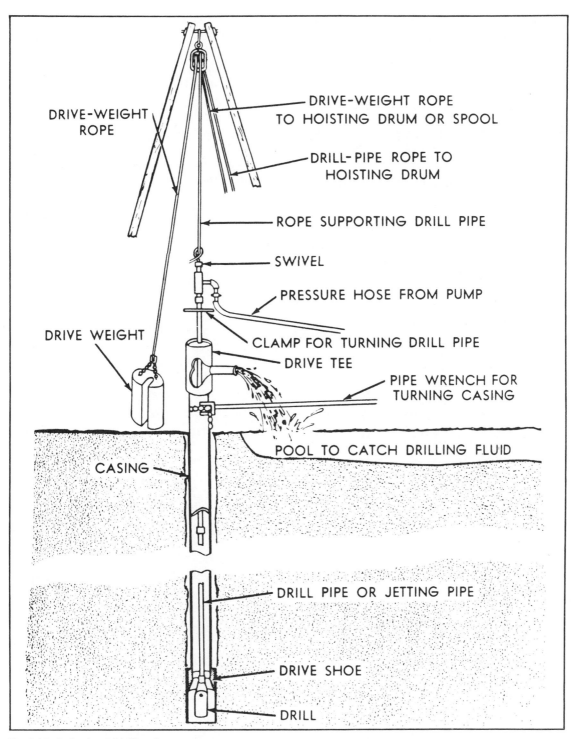

DRIVE-WEIGHT
ROPE

DRIVE-WEIGHT ROPE
TO HOISTING DRUM OR SPOOL

DRILL-PIPE ROPE TO
HOISTING DRUM

ROPE SUPPORTING DRILL PIPE

SWIVEL

PRESSURE HOSE FROM PUMP

DRIVE WEIGHT

CLAMP FOR TURNING DRILL PIPE

DRIVE TEE

PIPE WRENCH FOR
TURNING CASING

POOL TO CATCH DRILLING FLUID

CASING

DRILL PIPE OR JETTING PIPE

DRIVE SHOE

DRILL

Fig. 8-13. A simple, self-jetting rig.

feet with a hand auger. The casing is positioned vertically within the starting hole. Then the jet is positioned within the casing. Water is turned on and the casing is washed in. When the casing has reached the desired depth, the well can be completed as suggested. To facilitate casing penetration, teeth are often cut into the leading edge of the casing. In position, the casing is turned, as the water pressure is applied, with a chain wrench. The teeth act like a crude circular saw. See Fig. 8-12.

Driving a Casing. A drive shoe is fastened to what will be the lower end of the casing. This is a steel ring that has a sharp edge. A cap or a similar arrangement is positioned atop the casing to receive the strokes or the sledge of the falling weight. When water is to be used in concert with the driving force, a setup as shown in Fig. 8-13 can be used.

Other arrangements are also possible. Bear in mind that the force must strike the casing alone. The pipe leading water to the jet, sometimes called the *drill pipe*, must be free to descend as required and free to turn on its axis.

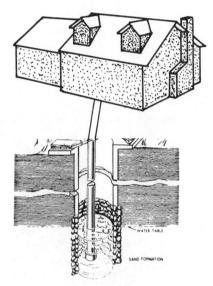

Chapter 9

Dug Wells

A DUG WELL DESIGN (FIG. 9-1) IS SIMPLE AND straightforward. It is simply a hole in the ground that is lined with brick, stone, tile, cement block, or even steel. The lower portion of the lining, the portion that is within the aquifer, is pierced so that water from the aquifer can enter. From the pierced section on up, the balance of the well wall is watertight.

Only water from the aquifer is permitted to enter the well, surface water and whatever might seep down from the surface or come through the zone of aeration is blocked. When the old oaken bucket was the only means of drawing water from a well, the walls of the well were brought up some 3 feet above the surface of the earth. A small roof was often built over the well opening and the rope on its roller was hung from the roof supports.

When the spout type of suction hand pump was developed, the walls were made almost flush with the surrounding soil and the pump was mounted on a slate or concrete cover over the well. Modern wells using pumps of one type or another can be terminated flush with the earth or even a distance below.

The well cap or cover can be then covered with earth and the well can be hidden from sight. See Fig. 9-2.

EVALUATION

If you want a romantic touch for your front or back yard with a shingled-well house with a rope and reel or a pitcher type of hand pump, the dug well could be your answer. If, however, you are seeking an economical way to secure portable water, the dug well is rarely the answer.

You probably have seen such wells and you probably know that they are tubular, watertight structures with openings to admit water at the bottom. If you have even a little experience with mortar and brick—or even if you have never held a mason's trowel in your hand—you can readily see that laying the brick, block, or stone that usually makes the walls of a dug well is no problem.

Laying brick or block or whatever within a hole in the ground is easy. The mortar and the brick provide the vertical strength. The circular design provides lateral stability, and this is backed up by

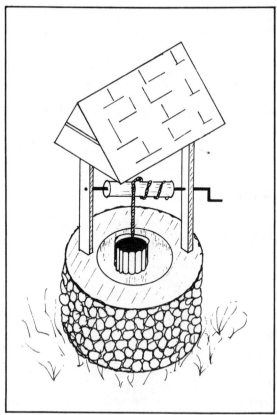

Fig. 9-1. The old dug well with its oaken bucket is still a common sight in rural districts.

the earth surrounding the structure. Neatness doesn't count because the work is out of sight. Unevenly laid bricks embedded in mortar provide as tight a water seal as bricks laid up with beautiful precision.

If you have the necessary pick and shovel, you don't need much more to mix up a little mortar and lay the brick or block. It would appear then, at this juncture, that all you need do is bend your back. This is the way they dug wells years ago and still do in many countries. And this is the way many well diggers have died.

You **CANNOT** safely enter a hole in the ground that is deeper than your waist unless the sides of the hole or the excavation have been shored up, curbed, or braced (the terms are used here interchangeably). There is no doubt a vast body of collected folklore describing the appearance of safe

and dangerous soil. Supposedly, there are soils that can be safely penetrated without shoring and soils that cannot.

Even in modern construction work, with all excavations thoroughly braced according to law, construction workers are killed every year when the walls of an excavation collapses upon them. The causes can be insufficient bracing, poorly position-ed bracing, or just bad judgment. Perhaps the brac-ing was removed too soon or someone began working in the bottom of the trench before the walls were shored up.

A dug well is not simply a lined hole in the ground. It is a small-scale engineering project. Un-less you take proper precautions, you could die constructing it.

Labor. The work consists of digging; that is obvious. What is not obvious is the quantity of soil that must be removed. It is far greater than can be easily visualized by just examining an existing dug well.

Obviously, the depth to which you will have to excavate is not known at this point. What you do know, or what you will soon decide, is the diameter of the projected well. See Fig. 9-3. Assume you decide upon a well with a 4-foot inside diameter. The excavation needed to accommodate the 4-foot well should be about 7 feet if you are going to use bricks for the well wall.

An excavation of this diameter will leave you about 1 foot of clearance between the brick and the shoring planks. If you build the wall of block, you have to increase, or you should increase, the exca-vation diameter accordingly. In projecting the ex-cavation figures. I am assuming that the walls of the excavation can be held perfectly vertical and rea-sonably smooth. Generally, this is not possible. The result is that you need move considerably more soil than merely the volume you might compute for a smooth bore hole.

The earth that you remove generally has to be hauled off somewhat. A little can be used to build a slight grade up to the wall of the well if it is to project above the earth. The soil that will not be required to fill the space left when the planks are pulled is left at hand.

The earth that is removed from the excavation

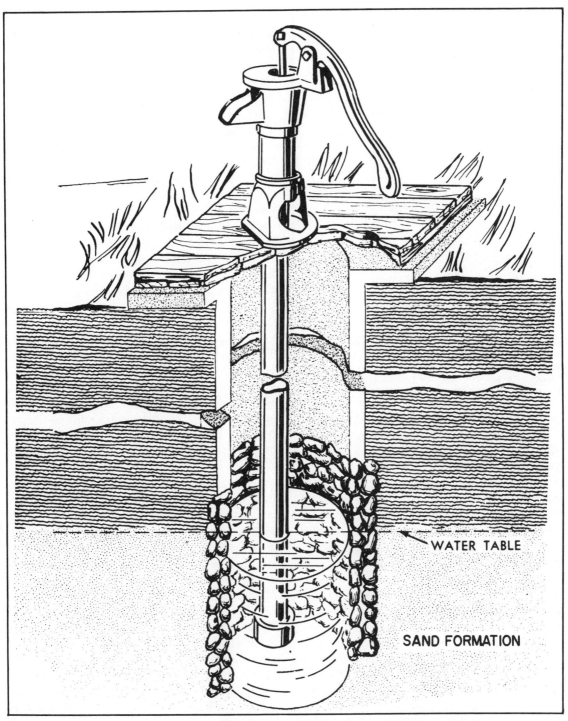

WATER TABLE

SAND FORMATION

Fig. 9-2. A dug well with "dry" stone walls in the aquifer and a poured concrete casing above. A pitcher-type hand-vacuum pump draws the water. Note how the concrete has been carried beyond the well to provide surface-water protection.

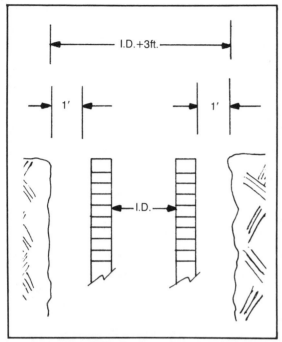

Fig. 9-3. When you dig the well hole, you must make its diameter sufficiently large to provide about 1 foot of clearance between the well-wall masonry and the surrounding earth. You need this space for working.

planks or a jack hammer with an attachment for driving piles—some type of tripod or derrick with which to lift and lower the hammer, an air compressor to power the hammer, and a high-power motor-driven pump and attached hose for draining the well when you reach water. That's not all.

First determine for certain that there is indeed water to be found and that it is both sufficient in quantity, suitable in quality, and at a depth within your reach.

Your practical, safe reach is no deeper than is possible with the length of your planks less the 1 foot that should project above the surface of the earth for safety. Wells have been dug more deeply than the length of a single plank. The method consists of installing a second set of planks within the first ring and driving the second set down into the earth as excavation proceeds. I *strongly* advise you to secure local engineering assistance before trying this.

Determine the aquifer depth before digging. This is most easily accomplished by driving or washing in any of the wells previously discussed: driven, jetted, and bored. You will need, at least temporarily, the parts and equipment required for installing any of the aforementioned wells. As you can see, the cost in labor and material is a lot greater than can be easily visualized by just looking at a completed dug well.

Advantages. Outside of its romantic appeal, there is only one advantage to the dug well, and that is its storage capacity. Whereas a small-diameter well—a well with a diameter in inches—can hold no more than a dozen gallons of water, a dug well with, for example, an internal diameter of 4 feet can have an at-rest water level that results in water to a depth of 10 feet and holds 940 gallons of water.

The equivalent would be a 4-foot, round tank, 10 feet long. That is quite a tank to purchase and store. The value of this is not only a reserve of cool water against temporary drought, but an immediate source of water for fighting fires and the like.

There are additional reasons why wells are not constructed in very small diameters even though an increase in diameter does not increase water yield to offset the great increase in well material and

has to be hauled up by hand—a bucket at a time. A little speed can be picked up by fastening a bucket to each end of the rope. This way when a full bucket comes up an empty bucket is lowered. Even so, moving tons of earth is a slow, laborious process when all you have is two hands a bucket and a rope. Bricks or blocks and mortar have to be lowered the same way. There is nothing fast about the method.

Material. You will of course need bricks or block and mortar. The exact quantity depends upon the diameter of the well and its depth. You will also need a quantity of size-graded gravel (if you pack the well) or ordinary gravel or crushed stone (if you do not pack the well). You will need 2-×-8-inch planks that are longer by at least 1 foot than the depth to which you will excavate. In addition, you will need either steel hoops or sufficient planks to form bracing rings of wood as shown in Fig. 9-4.

Equipment. In addition to the pick, shovel, ladder, bucket, and pulley (or whatever), you will need a maul—if you are going to hand drive the

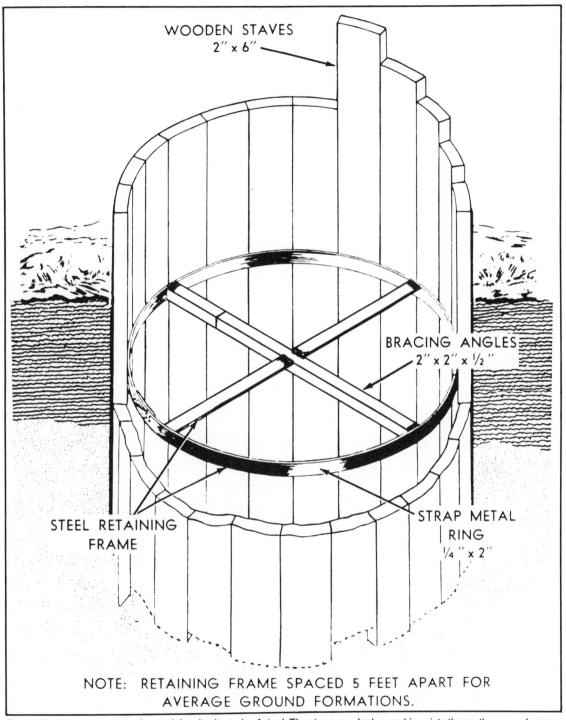

WOODEN STAVES
2″ x 6″

BRACING ANGLES
2″ x 2″ x ½″

STEEL RETAINING
FRAME

STRAP METAL
RING
¼″ x 2″

NOTE: RETAINING FRAME SPACED 5 FEET APART FOR
AVERAGE GROUND FORMATIONS.

Fig. 9-4. In this arrangement, the retaining ring is made of steel. The staves or planks are driven into the earth more or less one at a time.

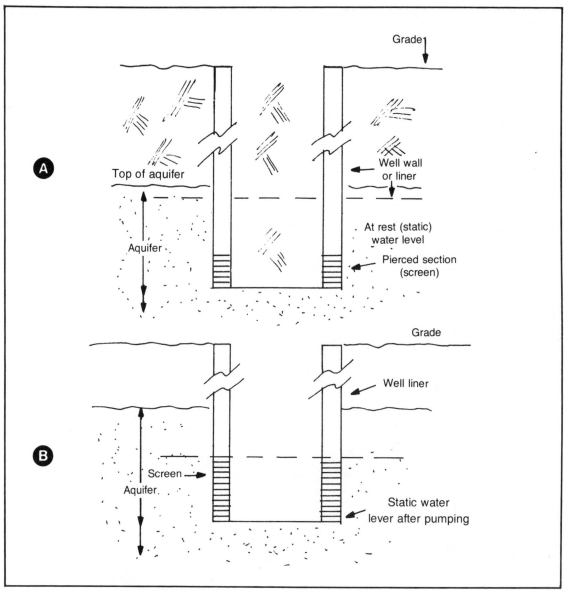

Fig. 9-5. The pierced section of the dug well wall should be placed sufficiently deep (if possible) within the aquifer so that pumping will not bring the static water level below the pierced section: A: shows a properly positioned pierced section before pumping; B: shows the same section still below the water level even after pumping.

labor. Very often a comparatively large well diameter is needed to accommodate in-well pumping equipment. For another, it is difficult to hold a very small-diameter, long well pipe perfectly straight. This is especially true when you are going through stratum of varying density.

A small-diameter pipe cannot support its own weight when it is very long. When the well has to go down to great depths—and the lower sections of pipe have to be lowered through the upper, in-place pipe—the uppermost length of pipe can have an inside diameter of a yard or more.

DECISIONS

Site Selection. The location of a well is selected with several factors in mind: the presence of water; relation to point of use (house or garden or barn); distance from probable or possible source of contamination; appearance of the well or well house if it is to terminate above ground; and convenience of soil removal and disposal.

Test Well. A small-diameter well is driven or bored (or whatever) to a depth that reaches the top of the aquifer, as determined by water flow, and then the well point is driven more deeply into the earth to determine if the aquifer is of a suitable thickness (or greater) for the proposed well.

Bear two things in mind:

☐ Initial well yield can usually be multiplied several times over (some report a nine-fold increase) with proper well development.

☐ Well yield is directly related to well screen height.

In the case of a dug well, yield would be directly proportional to the vertical pierced area height of the lower section of the well wall. Therefore, there is no point in digging a well 4 feet in diameter to install within an aquifer that is a foot or two thick. Figure 9-5 shows the proper placement of the pierced section of a dug well within an aquifer.

Well Diameter. The selection of the well's diameter should be based on the quantity of water you would like to store in the well. This means the water volume when the well is not pumped and the material you are going to use to line the bore hole. See Fig. 9-6.

If you plan to use common brick or fieldstone, there is no special advantage to making the well's diameter any particular size other than what is large enough to permit you or your assistant to work inside it. If you are going to use concrete block, you need to either use short block or to secure the sewer block (which is curved). In the latter case, the arc to which the block is curved controls or establishes, if you will, the inside diameter of the well. See Fig. 9-7.

If you are thinking of using precast, concrete sewer pipe for the well lining, bear in mind that you

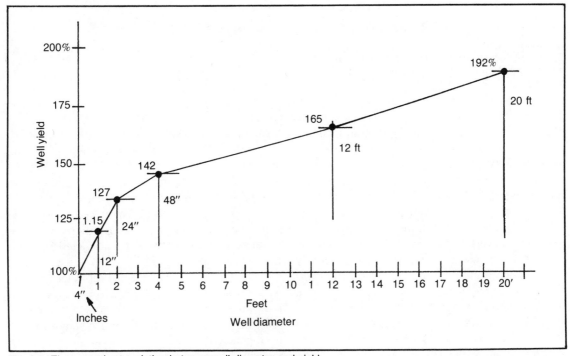

Fig. 9-6. The approximate relation between well diameter and yield.

93

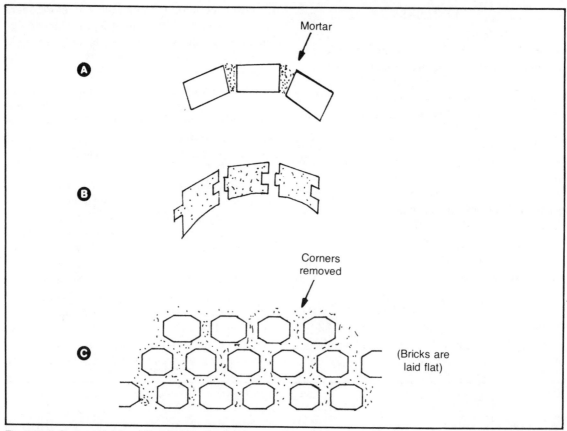

Fig. 9-7. A: Common brick can be laid flat to form a circle. B: Precast curved concrete block can be used. C: The corners of bricks should be removed to permit the passage of water.

will need some type of power equipment to lower the pipe sections into the well. Some of the concrete sewer pipe is manufactured with neoprene seals that automatically join the sections in a watertight connection.

Some sewer pipe requires that the pipe sections be mortared together. The use of this type of pipe means that you apply mortar to the upper edge, which should be the bell of the pipe, immediately before it is lowered into place. See Fig. 9-8. If you do go this route and hire a crane, you can probably have the crane pull the planks without extra cost. Generally, cranes and their operators are hired on the basis of time and distance to the job.

When you lower self-joining concrete pipe into the well hole, no one needs to be inside the excavation to complete the job as is necessary when

masonry is used for the well lining. If you are thinking of using a concrete pipe liner, consider hiring a large bucket borer to do the excavation. You will be limited to 24-inch pipe, but, if the aquifer warrants it, you can make up for reduced diameter by increasing the lining area within the aquifer—if the aquifer is thick enough.

Well Termination. You can terminate the well lining some 3 feet above the ground and leave the upper end open to the air but protected from rain by a small, shingled shed roof. If you are using concrete pipe, you can hide the top 3 feet of pipe with a veneer of brick or stone. You can terminate the well a few inches above the ground and seal its top with a masonry slab. Water can be drawn by a hand-pumped suction pump. You can also terminate the well several feet below the surface of the earth,

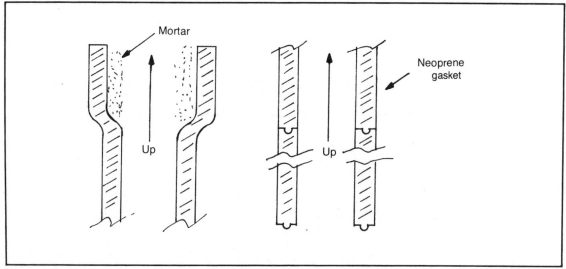

Fig. 9-8. Left: Mortar is placed in the bell of a precast concrete sewer pipe to seal the pipe-to-pipe spigot that will follow. Right: Neoprene gaskets seal these sections of pipe.

seal its top with a slab of masonry, and draw the water off with a pipe running sideways to the point of use.

You can use a suction pump here. Bear in mind the limitations imposed by altitude and the fact that there is always some loss of vacuum in any pipe. The longer the pipe and the more joints the more that is lost. On the other hand, there is plenty of room to install any type of pressure pump you prefer or require in the well itself. See Fig. 9-9.

To keep the pump noise out of the house and to make the pump more readily accessible for attention and service, the well can be terminated in a well house. This can be any type or design of small building large enough to hold the equipment and permit easy servicing and parts replacement.

The problem with a well house is frost, and this cannot be easily evaluated. Water freezes at 32°F and freezing is accompanied by an 11 percent increase in volume. Low temperature alone will not freeze water. It is a matter of how low, for how long, and what is the heat volume of the water. Insulation will delay freezing, but it will not prevent it.

While 4 or 6 inches of insulation will provide safe protection in one area, it will not be adequate in another area. To determine how much insulation

you need in your well house to protect the pipes and pump against frost, your best bet is to check with your neighbors and local builders.

Another strategy consists of providing a little permanent heat or some form of automatic, emergency heat. Building the well house against your existing home and cutting a little passage way for air movement between them will guard against freezing of the pump and pipes. Wrapping the pumps and pipes with electric heating wire and connecting the wire through a thermostat set to close at 40°F will also give you frost protection. The well house should be insulated or both the pump and pipes should be insulated if that is practical.

Still another solution to the frost problem is to dig down well below the frost line. This entails constructing a concrete enclosure with a heavy dirt and waterproof cover. See Fig. 9-10. The entire well termination enclosure must be below the frost line.

CONSTRUCTION

The planks are brought to the job site. One end of each plank is chamfered (angled), and the other end is rounded and bound with half a dozen turns of bailing wire. This is the end of the plank that will be pounded. You don't want to split or mash the plank

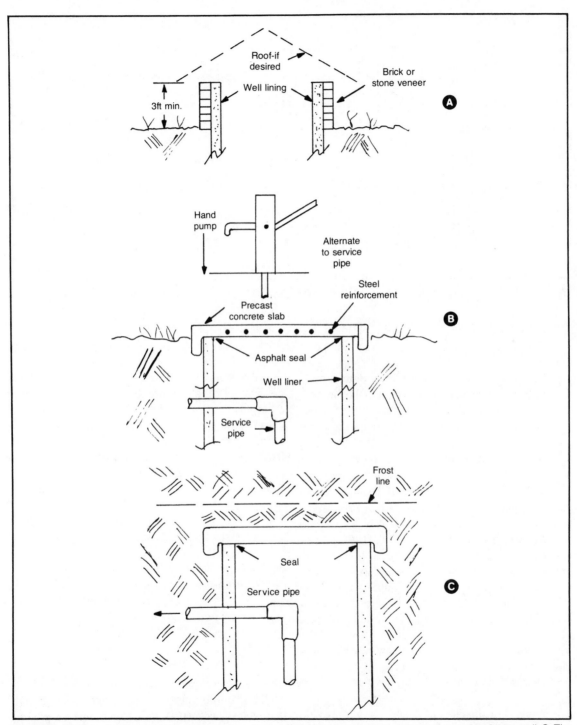

Fig. 9-9. A: Brick can be used to hide a well liner above the earth. B: A precast concrete slab can be used to close a well. C: The same cover arrangement used below grade.

end. About 1 foot down from the wired end, a 2-inch hole is bored through the plank. This will be of help when you have to pull the plank. When the planks are positioned, the chamfered edges are positioned inward, facing each other so that driving the planks tends to drive them outward against the earth. See Fig. 9-11.

Spruce is first choice for plank lumber, oak is next, and fir is last. Do not use poplar because it is too weak. Do not use fir with large knots because the planks will break at the knots. Use 2 or 3 × 6s or 8s.

You will also need steel rings or hoops exactly equal to the desired inside diameter of the wood

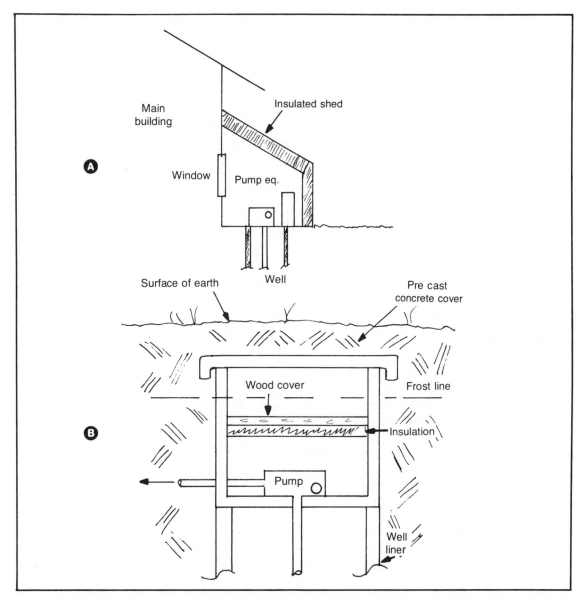

Fig. 9-10. A: A pump house constructed adjoining a building can receive light and heat through a window. B: The well pump is housed beneath the soil in a concrete structure and insulated from frost.

97

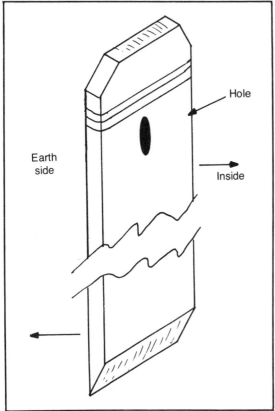

Fig. 9-11. A shoring plank that is to be driven has its top wrapped in wire to prevent splitting. The hole helps in pulling it up and the angle cut on the bottom tends to guide the plank's end against the soil.

form you are building. Remember that you want about 1 foot of clearance between the inside of the planks and the inside of the well lining. Because steel rings of this diameter are somewhat difficult to find these days, make your braces out of 2-×-8-inch planks. Use 14d nails to hold the parts together (as shown in Fig. 9-12).

Excavating. Dig about 3 feet into the earth. Take care to excavate the circle you want and to keep the bottom of the hole fairly level. Using a 2 × 4 and a couple of short boards nailed to its end, make yourself a gauge as shown in Fig. 9-13. This will make it much easier to keep from wandering from hole width.

Stand in the center of the excavation and dig a narrow trench around the inner perimeter of the excavation. Go just as deeply as you can. If the walls do not collapse, you can widen the trench a bit to gain more depth. The deeper you make this trench, the easier it will be to install the planks.

By leaving yourself an "island" of earth in the middle of the hole you are protecting yourself against cave-in. Even if the excavation walls should fall down on you, and there is no way anyone can predict this will or will not happen with certainty, you will be head and shoulders above the dirt and safe until you are dug out. See Fig. 9-14.

Position the planks vertically, in a circle, chamfered edges in the trench. Use baling wire and some nails to hold the planks together. Next, the planks are driven into the earth one at a time, a few inches at a time. Do not try to do this standing on a ladder, but do erect a high scaffold using some high horses and work from it.

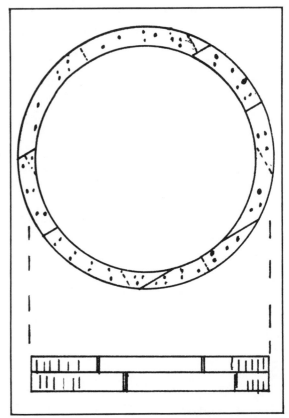

Fig. 9-12. A retaining ring made by joining a number of planks.

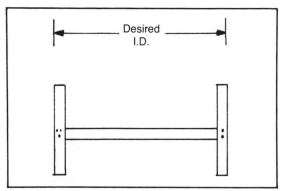

Fig. 9-13. A simple guide that will help you hold the well diameter to the desired dimension.

With the shoring firmly in place, you can remove the dirt island and extend the trench around the inner perimeter of the excavation. Between removing soil from beneath the ends of the planks and driving them, you work your way deeper and deeper into the earth. When you are down about 5 or 6 feet, it is time to install the circular braces. They do not need to be nailed to the planks. Spikes driven into the planks beneath the braces will hold them in the proper horizontal position. The planks tend to move inward and press against the braces. No permanent fasteners are required here.

Working in Water. When you reach the aquifer, water will seep or rush into the excavation. At this point your digging technique must change. Stop digging around the edges and begin digging in the middle of the excavation. You want to keep a hole about 1 foot or more deep in the middle. This is where you position the input end of the hose that is attached to the pump that will keep your well reasonably dry while you continue working. Make certain you have a dependable pump. If it suddenly quits and the low water in is great, you might get a bath—or even drown.

Constructing the Liner Base. If you are on gravel or very coarse sand when you finally reach the preferred depth—which should include at least one extra foot to accommodate the layer of fine gravel you will need to lay down to act as a filter to keep dirt and sand from rising into the well—you can probably lay your first course of brick or block or the end of a section of pipe directly down.

If you are on clay or silt or a mixture of clay and sand (or have any doubt), it is best to prepare a base or foundation for the well liner. The base (Fig. 9-15) can be constructed by simply laying down a circle of 3-inch-thick fieldstones or cut stones 1 foot square or larger. They are laid dry. As an alternative, you

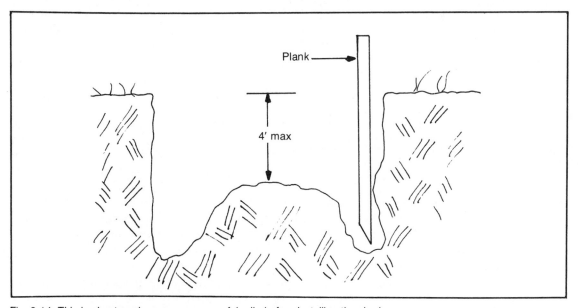

Fig. 9-14. This is about as deep as you can safely dig before installing the shoring.

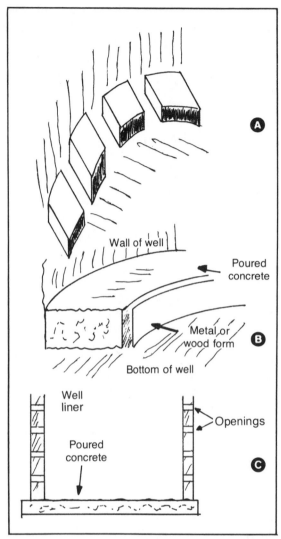

Fig. 9-15. There are several ways by which you can provide support for a well liner when the formation is very soft. A: You can lay down a circle of flat rocks. B: You can pour a concrete supporting ring. C: You can fill the bottom of the hole with several inches of concrete.

at this point despite the best efforts of your pump. No matter. Hold the form down with some rock and pour a thicker than usual mixture of concrete right into the water and into the form. The concrete, being heavier, will displace the water. The concrete will harden normally despite being covered with water. You will build or lay your well liner atop the foundation.

The Liner. The lower portion of the well liner is pierced by one method or another. It can be considered the well screen. Well yield is directly related to well screen length. Double the length (actually its height) and you roughly double the water quantity entering the well. This assumes that the screen is entirely within the aquifer. The upper portion of the well liner is watertight. You want water to enter through the screen and not seep down into the well from the surface of the earth.

When using bricks or blocks or stone, which are always laid flat, the corners are knocked off to leave openings. The masonry is laid up in mortar consisting of approximately 1 part cement plus 2 to 3 parts sand, plus ¼ part slaked lime. Or mortar cement that includes lime can be used.

The bricks or concrete block are laid up (positioned) in the standard manner. Take care to poke a stick between the corners of the adjoining units to keep the aforementioned openings open. Large, flat stones can be laid up dry; leave a little opening between adjoining stones. From the screen section on up, the masonry is laid up solidly in mortar.

In addition, the rear side of the liner, the side facing the earth, is given a ½-inch thick coat of plaster cement. This is the same mortar and sand mix previously described, but to which a little Anti-Hydros is added as a waterproofing agent.

Packing the Dug Well. It is important to pack the dug well. Packing is described in theory and practices more fully in Chapter 12. This section explains how it can be done with a dug well.

When liner construction has reached a point about 1 foot above the top edge of the screen section, work on the liner is temporarily halted. The space behind the screen, meaning between the screen and the soil, is filled with the packing mate-

can pour a concrete base. Construct a 6-inch high circle of wood or metal about 2 feet less in diameter than the inner diameter of the excavation. You can even make this out of a number of short boards; it doesn't have to be perfectly round. You just have to end up with a foundation 6 inches thick and about 1 foot wide in most places.

You might be working in several feet of water

rial. Ideally, this should consist of graduated gravel or crushed stone and coarse sand, positioned in three vertical layers. The layer next to the screen should be of stones about 1 inch across. They should be just large enough so as not to be able to easily slip through the holes in the screen.

These stones are followed by stones about ¼ inch across, and followed by coarse sand. See Fig. 9-16. No harm is done if the stones and sand become somewhat mixed. Don't waste time striving for perfection. You can use a sheet of galvanized iron, bent into a curve, to hold the stones separate while you are placing them.

With this done, the temporary spacer or spacers are removed. A 4- to 6-inch seal of concrete is now spread over the top of the packing. This is done to keep loose soil from working its way down into the packing. Now the balance of the liner can be constructed.

ALTERNATE TECHNIQUE

The technique just described for constructing a dug well is usually termed the *loose-stave method*. The staves or shoring planks are held loosely together by means of a wire wrapped around them. Each

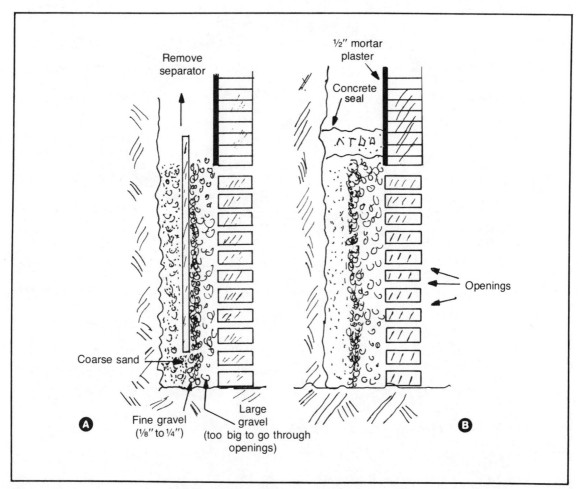

Fig. 9-16. The optimum method of artificially developing (packing) a dug well. (A). The temporary boards makes it possible to place relative coarse gravel adjoining the well and coarse sand between the gravel and the aquifer (B). Separators have been removed. The top of the packing has been sealed.

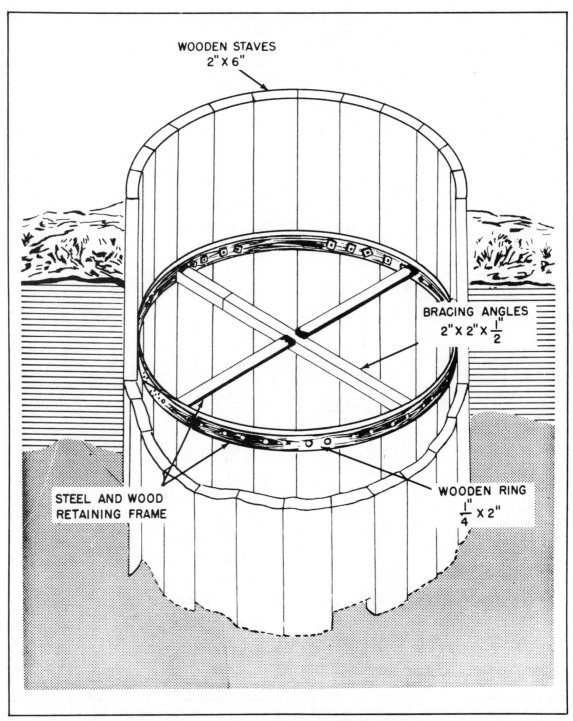

Fig. 9-17. This method of preventing the collapse of the walls of a dug well consists of building the entire bracing structure above ground and then lowering it as you dig.

plank is driven a few inches into the ground, one at a time. In the alternate technique, all the staves are strongly joined to one another and all move downward simultaneously. Additional alternatives consist of constructing the liner from sections of large-diameter concrete or clay pipe, or building a rigid liner on a steel shoe. The pipe or the cast masonry liner can also be "sunk" as a unit. See Figs. 9-17 and 9-18.

Sinking the Precast Concrete Pipe. The inside diameter of the cast pipe selected must be large enough to permit you or your assistant to work with pick and shovel while standing *inside* the pipe. The first section or the first two sections, which serve as the screen, must be pierced.

If you cannot secure the pipe with holes, you will have to drill them. One 1 or 1½-inch hole every horizontal foot and every vertical foot will be plenty. A clearance hole 3 or so feet deep is dug. It should have a diameter sufficient to provide 6 to 8 inches of clearance all around the pipe. The pipe section is now upended, bell section topmost, if it is to be joined to the following section with mortar, or slot down if neoprene gaskets are to be used to seal the pipe sections.

With pick and shovel, remove the soil from the center space beneath your feet. Then remove the soil beneath the edges of the pipe to a distance of 6 to 8 inches beyond the pipe. As you do this the pipe will sink down. With luck, it will not cock in the hole and you will not encounter boulders. When the upper edge of the first section of pipe is below the surface of the earth for a foot or two, the second section of pipe is positioned and sealed with mortar (if required). Digging continues until you have reached the desired depth. See Fig. 9-18.

To pack this design, drop stones—too large to easily pass through the holes in the screen—into the space between the pipe and the earth wall surrounding it.

Sinking Other Preformed Liners. For sinking preformed liners, the same procedure is followed using clay pipe and fixed staves. In the latter case, you will have to "load" the staves up with weights of some kind to force it to sink.

The approach to the preformed masonry liner installation is a bit different. Begin with a circular channel of steel having the desired inside diameter and a channel opening large enough to accommodate the width of your brick or block. The steel

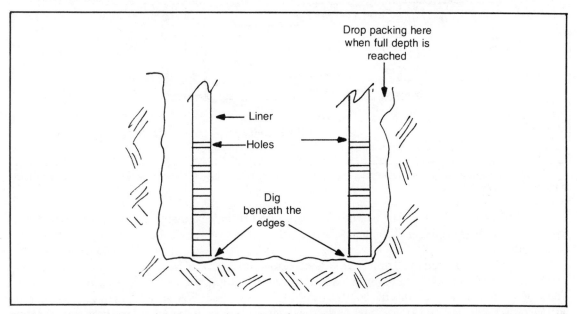

Fig. 9-18. This is the technique of digging beneath the edges of the well liner so as to lower the liner almost simultaneously with the digging of the well.

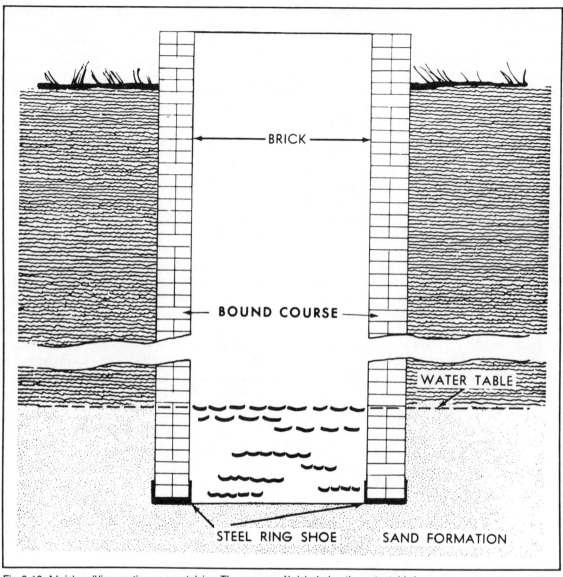

Fig. 9-19. A brick well liner resting on a metal ring. The courses of bricks below the water table have open spaces between them that are not shown.

circle, called a *shoe,* will have to be fabricated in an iron works unless you can find an old-time blacksmith.

The earth is excavated to a depth of 3 to 4 feet. The bottom of the excavation carefully made level. The shoe is now positioned within the hole, channel up. Next, with mortar and trowel you build the masonry liner atop the shoe to a height of 3 to 4 feet.

Bear in mind that you must leave openings for the water to enter. You can use stone for the filter section of the well lining, but you cannot lay the stone up dry. You must mortar them together because you must have a rigid structure. See Fig. 9-19. Now rest. Wait three days to give the mortar time to harden. If the air is dry, spray a little water on the masonry two or three times a day.

From here on down, it is dig and lay, dig and lay. You dig away beneath the steel shoe, as previously described, and you build up the liner as you go. Do not do too much on any one day because you depend upon your masonry work to hold back a cave-in and freshly laid brick work has almost no strength.

PROBLEMS

If you are going to depend upon a vacuum type of water pump (suction), remember that it is not very effective below a depth of 22 feet at sea level, and less so at altitude. You will need to keep the intake end of the pump hose in a hole below your final level. Remember to direct the pump outflow a considerable distance away from the excavation or the water will simply flow or seep back into the well hole.

Should you opt or need to use a gas-driven pump positioned within the well excavation, the exhaust must be carefully piped clear of the well excavation or the fumes could kill you. Even a slight leak in the gasoline engine exhaust system could do you in because the fumes are heavier than air and they will accumulate.

Caving. There is a chance you will experience caving even though you have shored the excavation walls properly. When you run into fine sand or water-bearing coarse sand, portions of the excavation wall "scab" off and fall to the bottom. There the soil turns to a kind of mud and piles up in the bottom of the hole. One solution is to fill the space between the earth and the shoring with gravel. The gravel is poured in gently and at a slow rate. When the gravel fills the space, the earth cannot easily cave-in.

Quicksand. When you reach the aquifer and have to pump, you will be excavating a drain hole for your pump intake nozzle to a depth a foot or more below the level of the soil on which you are standing. Should you strike quicksand, and you will recognize it by the ease with which your shovel penetrates, dig no further. You cannot rest the weight of your liner directly on quick sand. There is no telling how deeply it will sink.

At this point you have no way of knowing how thick the quicksand layer will be. If it is more than a couple of inches, it presents an insurmountable barrier. No matter how fast you dig it up, more sand will pour in from the sides, running almost like water.

You could either secure local engineering help or you could remove as much soil as you safely can from within the pipe. Then dig a number of channels beneath the pipe's edge. Pour 4 to 6 inches of concrete in the bottom of the well. With your shovel, force some of the concrete into the channels you have dug beneath the pipe edge.

COMPLETING THE WELL

If your well terminates in water-bearing sand or gravel, you can leave the bottom open. If it terminates in silt or clay, it is advisable to seal the bottom of the well. This can be done with a 6- to 12-inch layer of fine gravel or 4-inches of concrete. The purpose of the seal is to keep water movement from stiring up fine sand and silt.

If you have packed the well and topped the packing with a sealing ring of concrete, all you need do is fill the space between the liner and the soil with more soil—after removing the shoring. If you have built your well around a monolithic liner, it is advisable to pack the lower section of the annular space with stone of a size that will not easily pass through the holes in the screen section. These stones are followed with a layer of fine stones.

Atop the stones, pour a layer of concrete as previously described. When the concrete has set—a couple of hours will do it—the remaining annular space can be filled with soil. Wet the soil down to make it compact more easily and leave a quantity on hand to fill in the space when, with time, the soil subsides.

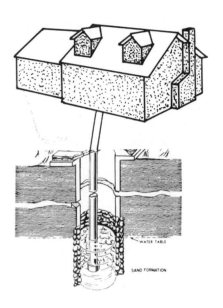

Chapter 10

Drilled Wells

WATER TABLE

SAND FORMATION

THE TWO MAJOR TYPES OF DRILLING RIGS (WELL-drilling equipment) used to reach underground water are *percussion cable tool drills* and *rotary drilling machines*. Percussion drills operate by means of a bit that is raised and dropped onto the earth to open the bore hole that will become the well when the casing and or the well pipe are installed. Rotary drilling rigs (Fig. 10-1) use a bit fastened to the end of a drill pipe to bore into the earth. The debris produced by the percussion tool is removed from the bore hole intermittently by means of a *bailer* (a type of pail). Debris produced by the rotary bit is driven out of the bore hole by air pressure, water, or mud.

General Comparison. *Cable-tool drilling*, also known as churn-drilling and percussion drilling, spudding and other names, is the much older of the two basic drilling methods. It is also the simpler method in that a cable-tool rig can be operated by one man when speed is not important. Rotary drilling requires three men for efficiency and safety on the job.

Cable tool is in many instances the slower of

the two methods because the work has to be stopped and the string, the tool bit, and attachments must be removed from the bore hole in order to clean out the hole with a bailer. In some formations, however, cable-tool drilling is the better method.

Because cable-tool drilling is simpler, the rigs cost less. In addition, their lower weight makes them easier to site in rough terrain. Another advantage is that cable drilling is accomplished without water, mud, or high-pressure air.

Variations. Although well-drilling equipment can be divided into two major categories, there are many subcategories, some of which combine techniques previously discussed in this book. The techniques or systems most often used are discussed in following paragraphs.

CABLE-TOOL DRILLING

The equipment used for cable-tool drilling varies from simple, homemade rigs not capable of penetrating the earth for more than a few dozen feet, and giant rigs that are mobile and self-contained. Their

Fig. 10-1. A trailer-mounted rig. Photo courtesy Ramco Inc.

basic principles of operation are identical. Except for greater control sophistication, their parts are very similar.

Cable-Tool Equipment. A platform or base, which might or might not be mounted on a truck bed, has a *derrick* or a *mast*, as it is usually called, that can be raised from a horizontal traveling position to a vertical position when the job site is reached. *Drawworks* is a generic term for the engine or motor and associated gears, clutch, and controls that operate the hoisting line (drawing pipe to the mast), plus the spudding equipment. This is an adjustable eccentric drive that alternately lifts and releases the cable holding the tools.

The cable that lifts and lowers the tools runs from a reel up to a sheave (pulley) at the top of the mast and then down to the tools, which consist of a "string." The cable is fastened to a socket that, in turn, is screwed fast to a drilling jar, and which, in turn, is screwed fast to a drill stem, and finally to the drill bit. See Fig. 10-2.

The jar can be best explained by describing it as a very loose joint. So loose that its two halves can separate for a few inches or more. When the cable is lifted, there is a hammer action upward against the top of the drilling jar.

The bit (Fig. 10-3) is the business end of the string of tools. The cutting end of the bit is flared out beyond the diameter of the bit's body in order to permit the bit to cut a hole larger than the diameter of the body and the string of tools that will follow it. For water well work, the bit will be several feet long, as will each part of the tool string to which it is attached. It is the weight of the string that provides the force that drives the bit into formation. It is the weight of the jarring tool that provides the impact that frees the bit when the spudding drive lifts the cable and pulls the entire string upward a short distance.

The bailer is essentially a long pipe fitted with a check valve of one design or another at its lower end. To use the bailer, the tool string is removed and the bailer is fastened to a second line, called a *sand line* or *bail line*, and is run from the drawworks up over the sheave at the head of the mast and dropped into the bore hole.

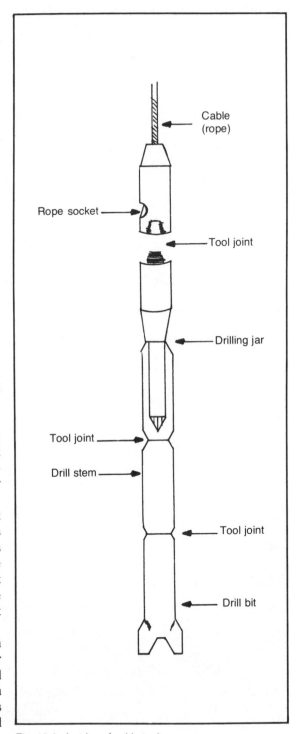

Fig. 10-2. A string of cable tools.

108

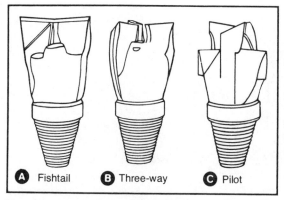

Fishtail   Three-way   Pilot

Fig. 10-3. Drill bits. Courtesy Koehler, Inc.

The weight of the bailer drives the lower end of the pipe into the cutting debris at the bottom of the bore hole. The bailer is then drawn up. As the bailer moves upward, the check valve automatically closes, locking the debris that has been driven into the pipe inside. Some drillers prefer to use a sand pump for cleaning out the bore hole in place of a bailer. A sand pump is a bailer with a rod and plunger inside to provide a suction that draws sand and small stones into the end of the bail pipe.

Bailing is continued until the bottom of the bore hole is reasonably clean. The cuttings are deposited clear of the job site and then removed. Bailing is very important because an accumulation of cuttings at the bottom of the bore hole cushions the impact of the falling bit, reducing its effectiveness to a considerable degree.

Drilling Procedure. Cable-tool drilling is always done with what the drillers call a "tight line." As they term it, the tool should be reaching for the bottom of the hole. This means that the release of the cable should be so timed that the cable is never slack as would be the case if the operator permitted the cable to unwind from its spool faster than it followed the bit into the bore hole.

When the cable (or rope as it is sometimes called) follows the tool, the cable always turns a little on its axis. This causes the bit to rotate and prevents the bit from striking the exact same spot in the hole with each stroke. A taut cable also saves time. No time is lost spooling up the slack. The bit

is pulled upward as soon as the jar reaches the limit of its freedom.

The second or two wasted reeling up slack that has developed in the cable might not seem to be important, but it is. *Drilling speed*, the rate at which the percussion bit is raised and lowered, can be as high as 60 times a minute. This is the responsibility of the drill operator. He has to adjust the spudding drive to the nature of the formation he is drilling and the action of the rope and tool string. This is done by varying both the speed of the stroke (how many times per minute the tool is lifted and dropped) and the length of the stroke (how high the tool is lifted before it is dropped).

The bit must be gauged very frequently. If its cutting diameter wears below a quarter of an inch, the bit will "freeze" in the bottom of the bore hole.

Bits can be sharpened in a portable forge at the job site. Some drillers retain worn bits for use farther down the hole.

As well depth increases, pipe diameter will be decreased every so often. When a casing cannot be driven any deeper, the only method of proceeding with the bore hole is to slip a smaller diameter casing into the in-place casing and continue on down. Wells that are expected to go thousands of feet into the earth will begin with an initial casing diameter of a yard or more (with a drill string and bit to match). Super depths are only achieved with rotary rigs. Percussions drilling is generally limited to a maximum depth of several hundred feet.

MUD-SCOW DRILLING

Mud scow drilling (Fig. 10-4) is a system of drilling that depends upon large, heavy bailers to produce relatively large-diameter holes in sand, clay, mixtures of quicksand and clay, or comparatively thin layers of quicksand and clay. The bailer used is similar in general design to the bailer previously described. The difference is that, in addition to its size and weight, there is the attachment of a drive shoe to its bottom end. A *drive shoe* is a sort of coupling or collar that is screwed or welded fast to the end of a mud-scow bailer or drill pipe. The end of the collar is sharpened to a knife edge to enable it to be driven into the formation more easily. The

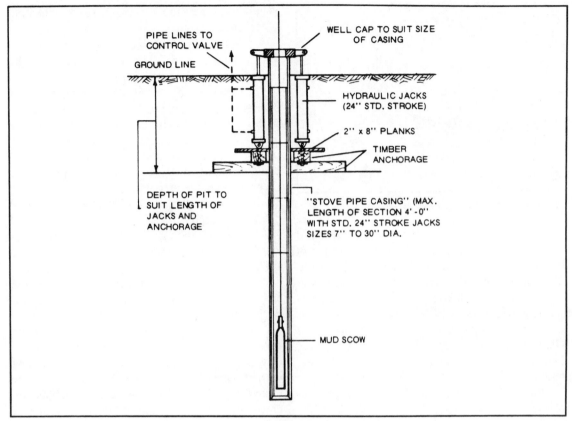

PIPE LINES TO
CONTROL VALVE

GROUND LINE

WELL CAP TO SUIT SIZE
OF CASING

HYDRAULIC JACKS
(24" STD. STROKE)

2" x 8" PLANKS

TIMBER
ANCHORAGE

DEPTH OF PIT TO
SUIT LENGTH OF
JACKS AND
ANCHORAGE

"STOVE PIPE CASING" (MAX.
LENGTH OF SECTION 4'-0"
WITH STD. 24" STROKE JACKS
SIZES 7" TO 30" DIA.

MUD SCOW

Fig. 10-4. Elements of a mud-scow drilling operation. Courtesy Koehler, Inc.

lower end of the mud-scow bailer is equipped with a self-closing valve, which is generally a plain, flat valve.

Mud-scow bailers are connected to jarring tools that permit longer jarring strokes. Stroke lengths for scows are usually on the order of 24 to 36 inches. That is considerably shorter than the strokes used with ordinary bailers. Mud scows are run at an average working speed of 25 to 40 strokes per minute. Stroke length depends upon the formation being drilled. In sand and gravel, the stroke is kept short. In clay the stroke can be much longer.

When you are percussion drilling through relatively stable formations—formations that will not close in on the bore hole—the well casing can be lowered into place after drilling has been completed. In caving formations, the casing follows the bit.

In mud-scow drilling, the end of the casing and the excavation of the bore hole proceed simultaneously. To be exact, the casing always precedes the working end of the scow by a short distance. Should the scow advance farther into the earth than the casing, there is a good likelihood that the scow will become jammed into the formation. The basic design or the basic purpose of using a scow in the first place is to drill into caving soil; that is something not easily accomplished with standard percussion drilling techniques.

Another major difference between mud-scow drilling and standard percussion techniques is that the mud-scow casing is "pulled" onto the formation and not driven.

Work at the job site starts with the excavation of a flat-bottom hole that is roughly square in outline. The hole will be 6 feet or deeper and 10 × 10 feet or larger as required. The excavation is centered on the selected bore-hole site. A shallow bore hole can now be excavated by hand to facilitate starting. Next, 3-×-8-inch timbers are laid down to form an anchor or skid to which hydraulic jacks, in a vertical position, are bolted. These are double-acting jacks; they can exert a pulling force as well as a pushing force. The excavation or pit is now filled with crushed stone or the like to hold the timber anchorage in place. The casing is positioned vertically and a well cap to suit the casing is placed on top.

The jacks are energized and exert a downward pull on the casing, driving it a distance into the earth. The mud scow is now put into operation within the casing. As the mud scow is lifted, dropped, and removed for emptying, as required, the jacks continue their downward pressure.

Because you are working in caving soil, the walls of the bore hole tend to collapse as the scow removes soil. The pressure of the hydraulic jacks forces the casing to advance downward into the space partially cleared by the action of the scow and partially produced by the collapsing formation.

JET DRILLING

Chapter 8 describes the basic principles of jetting a well. In that chapter simple, hand equipment and low-power equipment is covered. The type of jet drilling remains essentially unchanged. The only differences are the size or power of the equipment and unvarying use of percussion to drive the casing.

Commercial jet-well drilling is generally limited to 2- to 4-inch diameter wells (the inside diameter of the installed casing at its upper end), and is rarely driven to more than 100 feet. The equipment is comparatively simple and light. It consists of a mast, drawwork (power and gears, etc., that provides the spudding action and powers the spool or reel that lifts the casing into place), and a high-power, high-pressure water pump. Generally, all this plus tools, etc., are mounted on a light truck that is backed into place to start drilling.

In place of the solid bit used for standard percussion drilling, a hollow bit with a number of holes at its sides near the cutting edge is used. This bit is screwed fast to a drilling pipe and water at high pressure is forced down through the drill pipe and out through the holes in the bit. In the usual procedure, the casing is driven into the formation for a few feet, and then the drill pipe and bit is lowered into the casing. Now the drill pipe is lifted and dropped with short, fast strokes.

The drill bit chops up and otherwise loosens the formation. The high-speed flow of water out the holes in the bit washes the bit cuttings and loose sand up the annulus—the space between the drill pipe and the inside of the casing—and out of the bore hole. In sand, soft ground, gravel, and similar loose formations, this method of drilling is very fast. The casing either drops of its own weight or is driven to speed its descent.

Hollow-Rod Drilling. The hollow rod (Fig. 10-5) is a variation of percussion-jet drilling. This method differs in that no water pump is required and very little water is needed. Hollow-rod drilling is most effective and efficient in comparatively soft formations such as clay and sand. Usually, the technique is not used for wells over 4 inches in diameter.

Equipment consists of the standard mast, drawworks, spudding equipment, and associate parts. No water pump is required. Drilling is accomplished by a special bit at the end of the drill

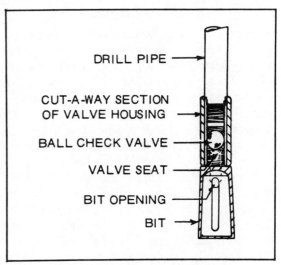

DRILL PIPE

CUT-A-WAY SECTION OF VALVE HOUSING

BALL CHECK VALVE

VALVE SEAT

BIT OPENING

BIT

Fig. 10-5. Hollow-rod drill. Courtesy Koehler, Inc.

pipe. There are openings and passages in the bit leading upward. The bit itself is mounted within a short length of tubing having a cutting edge at its lower end. The tube screws fast into a check valve that in turn is screwed fast to the end of the drill pipe.

In operation, the drill pipe and its bit is raised into a vertical position and dropped into a shallow, prepared hole in the earth. The drill pipe, which with its special bit is now called a *hollow-rod drill* is dropped, raised, and dropped.

The hole is filled with clean water and kept filled. The action of the drill bit cuts into the formation and the drill cuttings and loosened sand and soil form a slurry with the ever-present water.

As the hollow-rod drill is raised and lowered, a portion of the slurry is forced into the openings in the bit, up the bit and past the check valve, and up into the hollow-drill pipe.

As you can see, the up and down action of the bit and associated parts act as a sort of a pump forcing the slurry into the drill pipe. When the pipe is partially filled with slurry, it is completely removed from the bore hole, laid on its side, and tipped over to remove the slurry. This is the method often selected for drilling wells up to 4 inches in diameter in soft formations of sand and or clay. See Fig. 10-6.

ROTARY DRILLING

Power boring is described in Chapter 7. In essence, an auger is revolved by means of a power drive. The vertically positioned auger cuts a hole into the earth. The technique of rotary drilling is somewhat similar, and also somewhat similar to that used for jet drilling. The working end of the drill consists of a bit pierced by a number of holes that lead to the hollow interior of the bit. In turn, this leads to the hollow interior of the drill pipe. The drill pipe is fastened to a fluted section that passes through a shaped opening in a flat, round horizontal "table." In operation, the drill pipe and string of tools are assembled in a vertical position with all of the drawworks and the mast. See Figs. 10-7, 10-8, and 10-9. The kelly, a shaped bar, is passed through the similarly shaped hole in the table. Then the entire string is lowered until the bit rests on the formation. When the drive engine or motor drives the table, it rotates and this rotates the kelly along with it. Mud is pumped down the drill string and out through the holes in the bit.

As the bit turns, it chews into the formation and produces cuttings that are brought to the surface by the continuous flow of mud. The mud emerging from the annulus is directed to a settling pit. There the particles of stone and coarse sand settle out of the mud and clean, filtered mud is withdrawn and pumped back down the center of the drill string.

Drill-bit design varies with the type of formation expected or actually encountered. Simple fishtail or star-shaped bits are used in soft formations. Bits with short tooth cutters are used for hard formations. For working formations with medium hardness, cone-type and roller-type bits are used. For very hard-rock formations, bits with carbide buttons are used.

Soft-formation bits are usually rotated at a speed of 50 to 150 rpm. For hard formations, rotary speeds are reduced to a range of 30 to 50 rpm. The weight or pressure on the bit is produced by the bit and drill string, and often by the addition of an extra heavy section of drill pipe. This is done so that at no time is the drill string permitted to stand free of the attached cable. Should it do so, the drill string would probably bend and begin to drill at an angle instead of perfectly vertically as preferred. Weight applied to the drill bit varies from 2000 pounds to 5000 pounds per square inch of bit diameter. The lesser weights are used with soft-formation bits.

Large Diameter Holes. Bore holes having a diameter of more than a foot or so are usually drilled in stages. First a small-diameter bore hole is drilled into the formation, and then the hole diameter is enlarged with a roller bit. In other words, large diameter holes are first opened with a pilot drill.

When it is necessary to enlarge a hole directly beneath a casing so that the casing can be dropped or driven down, one of two types of underreaming tools are used. One depends upon hydraulic power. A fluid under high pressure is jetted through holes in the underreaming tool and against the walls of the

Fig. 10-6. Air operated, rotary-table rig. The rig is owned and operated by Petty Ray. Photo courtesy Holemaster, Inc.

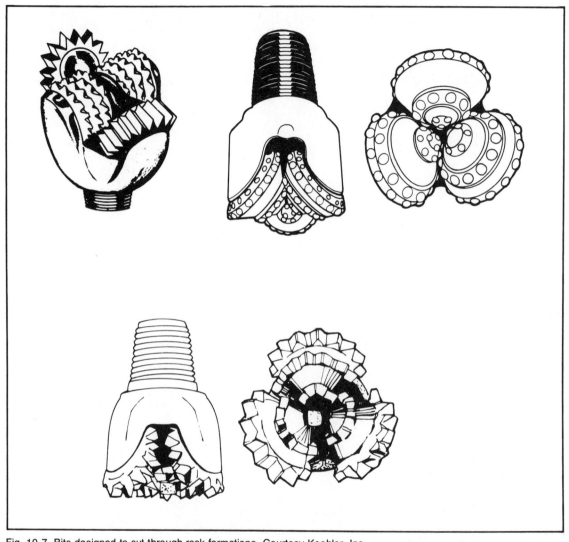

Fig. 10-7. Bits designed to cut through rock formations. Courtesy Koehler, Inc.

bore hole. The other underreaming tool has pivoting blades that spring outward when hydraulic pressure is applied to the underreaming tool.

The Mud. Sometimes just called plain mud and sometimes called the drilling fluid, it is rarely simple mud and water. More often it is a mixture of various chemicals and substances such as Bentonite, a special clay. See Fig. 10-10.

The drilling fluid performs a number of important functions: it removes the results of the bit's cutting action; it cools the bit to some degree; and it

prevents the walls of the bore hole from collapsing upon the drill string. Pressure by the fluid on the walls of the hole helps to keep the wall from collapsing. At the same time, the drilling fluid can also produce problems at the lower end of the bore hole. This is the area in which the well screen will be positioned.

The turning of the drill string and the pressure on the fluid acts to produce a firm layer or smear of clay on the bore-hole wall. The problem is to control the drilling fluid so that proper sealing can be

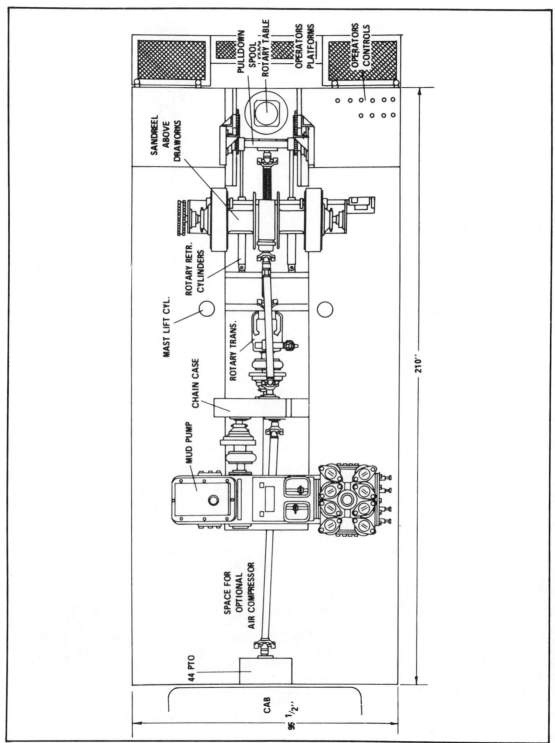

Fig. 10-8. Principal parts of an SS-10 Speedstar drill rig. Courtesy Koehler, Inc.

115

Fig. 10-9. A light rig owned and operated by Stamm Schelle. Photo courtesy G. E. Failing, Inc.

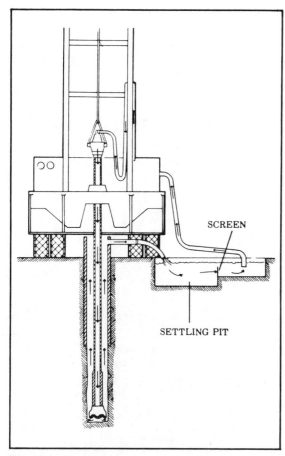

Fig. 10-10. Using mud as a drilling aid.

accomplished without the mud penetrating the bore-hole walls to such a degree that is later difficult to develop the well. The well wall is sealed with mud that naturally stops or slows the entrance of water until the mud is removed by developing.

Experienced drillers maintain that mud control is often the most important aspect of rotary drilling. Mud balance, meaning the weight per gallon of drilling fluid, is watched and generally held to 9 pounds per gallon. The pH of the mud is held to 8.0 to 9.0. Sand content is kept below 2-3 percent.

AIR-ROTARY DRILLING

An air-rotary drilling rig (Fig. 10-11) consists of the usual mast and drawworks, but has no drive motor or engine coupled to a rotary table that rotates the drill pipe by means of the kelly, the fluted or square or hexagonal bar that passes through the kelly. Instead, an air-rotary drilling rig will have a motor or an engine driving a compressor. Air from the compressor will drive the rotary table. Such an arrangement is further classified as an air-rotary, table-driven rig.

Alternate Designs. When the drill string is not rotated by means of a rotating table and a kelly—when the drill string is rotated by a slide-mounted motor atop the drill string and secured to the mast—the rig is called a top-headed drive rig. The slide arrangement permits the drive motor to remain attached to the mast and still follow the descent of the drill string.

A variation on both designs consists of the addition of a hydraulic pump and the replacement of the aforementioned air-drive motors with hydraulic motors.

Further Application of Air Power. Whether the drill string is rotated by one mechanical system or another—by hydraulic motors or air motors—all air-rotary drilling is accomplished with a constant stream of high-pressure air driven down the drill string and out through holes in the bit. The rapidly moving air clears the face and cutting surface of the formation of debris and carries the debris up through the annulus to the surface of the earth.

The advantages of the air system over the mud system is that no water is needed. No mud is used so no layer of water-sealing clay forms on the surface of the bore hole. The disadvantage of the air rotary system is that it can only be used in noncaving formations, or in caving formations with the simultaneous inclusion of a casing.

The roller type of rock bits are usually used when drilling in hard formations, with shorter teeth for very hard rock and longer-toothed bits used with soft formations. Rotational speed ranges from 10 to 20 rpm on very hard rock. With softer rock, the rotational speed is increased. Down pressure also is adjusted to the relative hardness of the formation through which the bit is cutting. The softer the formation, generally, the less down pressure is applied. Harder formations call for greater down pressures.

117

Fig. 10-11. Using the "pull down" to aid drill penetration. This is chain belt seen to either side of the kelly. Rig owned and operated by CON-FLEX. Photo courtesy G. E. Failing. Inc.

118

AIR-HAMMER DRILLING

Air-rotary drilling is a straightforward modification of the basic rotary-drilling technique. The major difference is the replacement of liquid as a drilling medium with air. Air-hammer drilling takes this change and carries it one step further. The old churn or rotational hollow-drill pipe is still used, but the drill pipe or string is not simply rotated to the cutting action by the drill bit on the bottom end of the pipe. With the air-hammer method, the drill bit is retained, but it is also pounded repeatedly.

Described very simply, a jackhammer has been placed at the lower end of the drill pipe. The pipe turns and the jackhammer pounds the bit into the rock. This is currently the fastest method of penetrating hard formations.

Simply pounding on the bit would eventually "lock" the bit into its hole. That is why simply lowering an air hammer or jackhammer to the bottom of the bore hole would not work. Rotation keeps the bit from striking the very same spot again and again. Rotation helps keep the bore hole vertical and the weight of the drill pipe adds pressure to the bit's cutting edge.

The air hammer attached to the bottom of the drill pipe consists mainly of a piston (Fig. 10-12), one or more valves, and the cylinder in which the piston moves up and down. Although air-hammer designs vary, they all operate the same way. Air under pressure comes down the drill pipe, enters the cylinder through a valve, and drives the piston down. The piston strikes the top of the bit. The bit is driven down. A valve redirects the flow of air to the underside of the piston. The piston is raised, ready for the following cycle. Expelled air is directed at the face of the cutting edge of the bit. The air scours the bottom of the bore hole and carries the debris of the cutting up the annulus and out of the well hole.

To facilitate changes and refacing, all the bits used with air hammers are interchangeable. This permits the driller to remove a drill bit, shape it, and possibly re-temper it without losing drilling time. Bit faces or cutting surfaces are not alike. Some have two or more air holes in the face of the bit. See Fig. 10-13. The air from these holes blows directly across the bit's cutting edges and onto the surface being drilled. In other designs, the air holes are at the sides of the bit. In all the designs, the air blows chips away from the bottom of the bore hole and up the annulus.

Depth Limitations. In a dry hole, drilling depth is limited by the weight the mast and associated reels or spools can carry. One technical description of a drilling rig (any type rig) is the mast hook capacity. On a really large rig, such as used for oil drilling, this figure can run to 1.5 million pounds. A comparatively small rig such as used for water well drilling of moderate depth might typically have

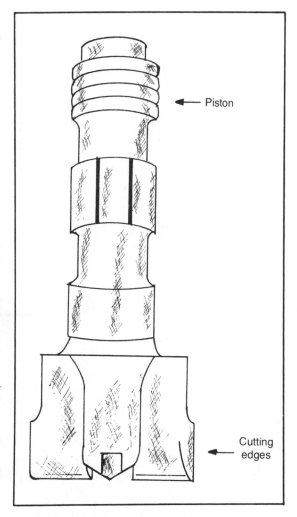

Fig. 10-12. The active portion of a down-hole air hammer. Bit and piston are one.

119

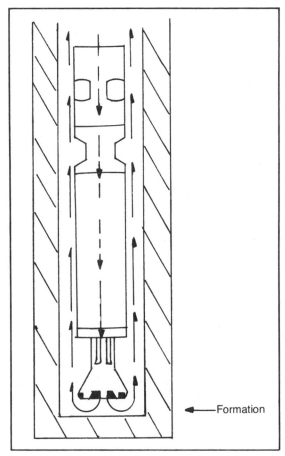

Fig. 10-13. In-hole air hammer showing air flow.

The longer the drill pipe the thicker it must be to sustain its own weight and to transmit rotational power to the working end. Safe drilling depth can be easily calculated by comparing the total expected weight of the bit and drill pipe to the total safe capacity of the draw motor.

Using the above guidelines oil drillers have gone down miles with the use of a drilling fluid. Air drilling, whether simple rotary or rotary plus in-hole air-hammer drilling, is limited by the presence of water. As long as the hole is dry, only the weight of the drill string versus the capacity of the rig's power limits drilling depth. When the bore hole fills with water, the level of the water in the hole limits the operation of the piston.

The deeper you go the greater the quantity of water that collects in the bore hole. The absolute limit is reached when the pressure of the water column equals the pressure of the air at the air hammer. That is the theory. In practice, this condition and associated depth is never reached because air-hammer efficiency falls off rapidly. Too much energy is wasted just driving the piston against the pressure of the water, too much energy is lost driving the bit through the water, and too much energy is wasted driving rock chips and water up through the annulus.

Air-hammer drills are usually rotated at speeds of 5 to 35 rpm, depending upon the hardness of the formation. The harder the rocks the slower the rotational speed. The reciprocating speed of the bit itself is typically 60 cycles per minute. The exact speed varies with down-hole air hammer design, air pressure, bit size, and the characteristics of the formation. The harder the rock the faster the bit rebounds.

Typically, air pressure for standard, air-rotary drilling varies from 30 to 50 psi. For air-hammer drilling, the required air pressure is considerably higher and ranges from 125 to 250 psi.

Air Power Requirements. Air pressure needs vary with the type of air-rotary drilling employed. Pressure needs also vary with the depth to which the driller plans to go; the deeper the projected hole the greater the required pressure.

There is an important relationship between

a maximum mast hook capacity of only 24,500 pounds (sometimes called hoisting capacity).

Even so, the smaller, truck-mounted rig can go down to 1000 feet or more with a 10¾-inch hole in overburden (moderately hard formations) and the same depth with an 8-inch hole in hard-rock formations. The above are typical figures for an air-rotary drilling rig. Using mud for drilling, the same (or almost the same) rig, in the sense of available power, can put down a 12-inch hole to the same 1000 feet or more.

A 1000-foot depth limitation affects all types of rotary-drilling rigs. The power of the drawworks is limited in lifting the drill pipe out of the bore hole. The casing is rarely removed once it has been positioned within the bore hole.

bore-hole diameter, drill-pipe diameter, and required air flow. As shown in Fig. 10-14, pipe size and hole diameter determine required minimum rate of air flow. This is generally accepted as being 4000 ft/min or something less than 80 mph.

Assume you have a specific hole diameter, let's say, of 5½ inches. Follow the line shown in Fig. 10-14 until it reaches a curved line sweeping upwards. These lines represent the size of the pipe that will deliver air to the bottom of the hole. In this example, it is point A on the 5-inch pipe. Now go right from point A to point B. From B, you read up to find a required air volume of about 450 cfm (cubic feet of air per minute).

If you want air to blow out of the annulus around 5-inch pipe in a 6½-inch hole at a velocity of close to 500 ft/min, you have got to drive air down the air pipe at a rate of 450 cfm.

For maximum air efficiency, the drill-pipe size should be that shown in Fig. 10-14. For example, if you plan to drill a 6½-inch hole, best efficiency would be obtained by using a 5-inch drill pipe. That would require 450 cfm (cubic feet of air per minute) of air passing through the annulus, which would mean the air traveled at a rate of nearly 80 mph. If you elected to use a 3-inch drill pipe in the same 6½-inch hole, you would require about 700 cfm of air. In Fig. 10-14, the lines sloping from left to right and marked with inches are drill-pipe-size lines.

Assume you plan to use a 5-inch drill pipe in a 6½-inch hole and, you need a constant flow of 450

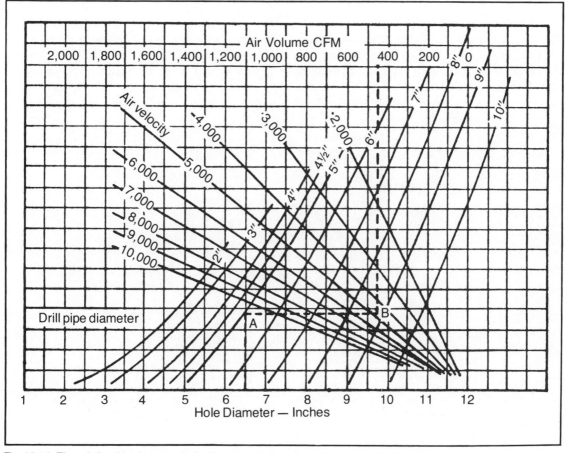

Fig. 10-14. The relationships between hole diameter, air-pipe size, air volume and air velocity.

cfm of air or more. The selection of a compressor and its accompanying drive also depends on the pressure at which this air must be delivered. Simply to specify a compressor of a particular air capacity is only half the specifications. You also need air pressure.

For standard rotary-air drilling, you would require a compressor capable of supplying air at a volume (in this example) of 450 cfm of air at a pressure of 30 to 50 psi. For a rotary air-drilling rig with the same setup, you would need a compressor capable of delivering the same volume of air, but at a pressure of 125 to 250 psi. Going from a requirement of 50 psi to 250 psi at the same volume would require a roughly corresponding increase in compressor power.

REVERSE-CIRCULATION DRILLING

Some drillers consider reverse-circulation drilling a form of excavation. The technique is best suited to sand and gravel formations. It is not suited to any formation much harder than soft rock. In sand and gravel, it is the fastest method of opening a large-diameter well hole. Hole diameters that are feasible at this time range from a minimum of 10 inches to a maximum of 60 and more inches. When drilling the 10-inch or slightly larger diameter hole, efficient drilling has reached to depths of 1500 feet.

Using 6-inch flanged pipe, reverse-circulation drill rigs (Figs. 10-15 and 10-16) have gone down to 600 feet with bore-hole diameters from 18 to 60 inches. Reverse drilling is usually chosen to drill large-diameter holes through a soft formation where the chance of cave-ins are very possible. The technique can handle boulders up to 6 inches across that might be encountered or might fall into the bore hole. The ultrahigh speed of the water utilized as a bore-hole cleaning means is capable of lifting stones that large.

Principle of Operation. Standard rotary-drilling systems depend upon a liquid or air to clean the bottom of the well hole. In standard systems, the liquid, mud, or air is driven *down* through the drill pipe and out through holes at or near the cutting end of the bit. The conveying medium deposits its load outside of the bore hole on the ground.

When water or mud is used, the mud and or water are channeled off to a settling pit and given

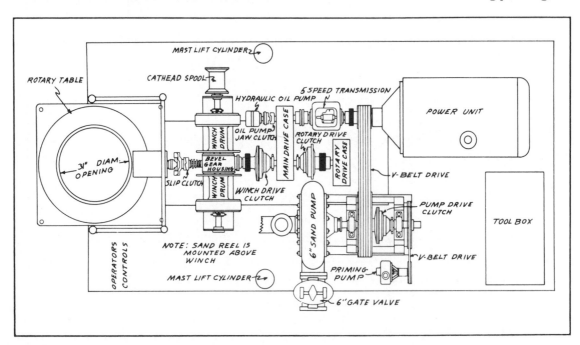

Fig. 10-15. Major parts of a reverse circulation rig. Courtesy Koehler, Inc.

Fig. 10-16. A reverse circulation rig owned and operated by the Rowe Bros., Inc. Photo courtesy G. E. Failing., Inc.

time to separate. The relative clear liquid or mud is then pumped back down the drive pipe. If air is the carrier, the air loses its speed as soon as it leaves the well hole and deposits the chips and dust alongside the well hole.

In reverse-circulation drilling, the carrying medium is directed *down* the annulus and is sucked *up* through the holes in the bit and on up through the drill pipe to the surface (very much like a vacuum cleaner).

In reverse-circulation, heavy mud is never used—just water. The walls of the well hole are not coated and sealed with mud that will stop or reduce the inflow of water to the well. Whereas a large-diameter well drilled by conventional rotary methods could require weeks to complete, reverse-rotary drilling can accomplish the same end in a fraction of the time.

The quantity of water required is considerable. Roughly, three times more water is needed than necessary to fill the planned hole to its top. This water is kept in a settling pit. There the solids drop to the bottom and the clear water can be removed by the pump and sent back down the bore hole. On an average well, water at the rate of about 1000 gpm (gallons per minute) is circulated.

This water speed does not overly disturb the walls of the bore hole. Because the volume of the annulus is considerably larger than the volume of the drill pipe, water traveling down through the annulus travels at a speed of roughly 1/20 that of the water ascending in the drill pipe.

Recently, the reverse circulation technique has been improved with the addition of compressed air to the power driving the circulating water. The air pressure drives the drilling fluid even faster than is possible with the pump alone. The increased fluid velocity results in faster bottom-hole scouring, faster clearing of blockages caused by loose formations falling into the well hole, and considerably deeper well penetration. Rates of up to an average of 40 feet per hour have been achieved.

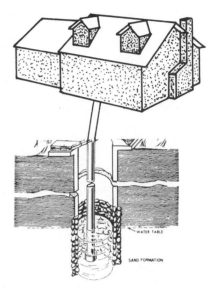

Chapter 11

Well Screens

A WELL SCREEN IS SOMETIMES CALLED THE "BUSI-ness end" of a well because of its importance in the efficient performance of a well. As a screen, it must prevent unconsolidated material such as sand from entering the well pipe, reaching and damaging the pump, and interfering with the use of the water for domestic and industrial purposes. At the same time, the screen must not interfere with the flow of water, which would reduce head pressure. While it is screening out sand, the screen also acts as a structural element in the bore hole to retain unconsolidated bore-hole wall material. A well screen acts as a screen and it is a section of bore-hole casing.

There are many companies manufacturing well screens to industry-adopted standards. Such screens are more expensive than homemade alternatives such as slotted pipes, but the difference in performance and lifetime more than makes up for the difference in initial cost. Properly designed and constructed well screens have continuous, uninterrupted slots circumscribing the screen's circumference. There is a minimum of solid space between the slots, leaving a maximum of open space for the passage of water. The slots have a V shape so that the smallest opening is at the exterior surface of the slot and the slot opening increases towards the center of the screen.

Only one metal is used in the construction of the screen. This virtually eliminates galvanic corrosion in the screen itself, but it does not prohibit similar corrosion between the screen and a well pipe that is constructed of a different metal. The screen must have sufficient physical strength to resist deformation in handling and placing and to resist the pressure of possibly caving bore-hole walls. The screen should be provided with an end fitting to make the screen adaptable for a wide variety of installation and operating conditions.

Types of Construction. What appears to be the best screen design for general purposes is the continuous-slot screen made by wrapping cold-drawn wire with a triangular cross section around a number of rods arranged in a circle. The triangular wire makes contact with the rods along its knife or thin edge. All these contact points are then joined permanently to the rods by welding. This results in a single, one-piece assembly that is very rigid and very strong.

The metals most often used are Everdur, silicon red, red brass, Monel, type 304 or 316 stainless steel, galvanized Armco iron, or galvanized low-carbon steel. For especially corrosive waters—which, of course, would not be drinking water—other, special alloys are used. In the last few years, plastic has also entered the well screen field. The material is strong and long-lasting; it will not corrode.

The opening-to-wall area of the screen is very low. Actually, the openings are no more than a series of horizontal slots usually spaced farther apart than the width of each slot opening. Where the aquifer is generously supplied with water and the need for a high draw rate is low, plastic screen (and plastic well pipes) become practical and economical.

Another, but less desirable, method of constructing a continuous slot screen does not employ welding to fasten the triangular wire to the vertical rods. Instead, the wire is forced into slots or notches cut in the rods or bars. The result is a savings in manufacturing cost at the loss of screen rigidity and screen strength.

The continuous-slot, welded-in-place well screen can be readily manufactured with almost any opening or spacing between turns of the wire. If a slot opening of 0.090 is called for, the successive turns are spaced that far apart. If a narrower opening is preferred, such as 0.020, the rolling and welding equipment can be set to this figure.

If there is a need for a change in openings between the lower and the upper end of the screen or even between the middle and the ends of the screen, this too can be accomplished during fabrication without difficulty. The need for varying slot spacing in a single screen might appear unnecessary, and it is in a 5-foot long screen or even a 10-foot long screen. But some screens in commercial or industrial installations can run to 30 and more feet, with outside diameters of 36 inches.

The value of the triangular cross section of the wire—which, as welded in place, results in slot openings that are smaller at the surface of the screen than they are at the inside of the screen—is that these slot openings do not clog as readily as would parallel-sided openings or openings that were larger at the outside of the screen than on the inner side of the screen.

When a slot has the narrow opening as the

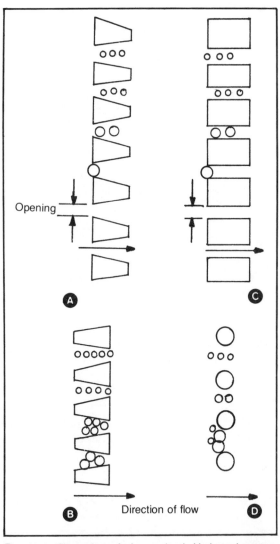

Fig. 11-1. The value of sharp-edged, V-shaped screen openings. A: If a grain of sand can get past the opening it will go all the way. B: If a grain of sand has difficulty passing through the slot it very likely will be jammed there. C: If the V slots are reversed, clogging and jamming is quite probable. D: Spaces between round bars act just as spaces shown in C.

126

sharp edges these slots have, a grain of sand that passes through the opening easily passes on. When the slot has parallel sides, a marginal-size grain of sand has to squeeze past the entire length of the slot's walls to get through the screen and travel beyond it. The chance of the grain of sand lodging in the slot is therefore much greater. When the slot opening decreases in size when you pass from the exterior of the screen to its interior, the slot becomes a trap for all grains small enough to enter the entrance to the slot, but too large to pass all the way through. See Fig. 11-1.

Developing a well depends on passing the fine grains of sand and silt through the openings in the screen. This cannot be accomplished satisfactorily should any particles become clogged in the screen. The continuous slot type of well screen has only two possible contact points that will impinge upon the movement of a particle and into the well pipe. For all practical purposes, the continuous, triangle-shaped wire screen is nonclogging.

Efficiency. Well screen efficiency is measured by the ratio of intake area per square foot of screen surface and water intake velocity. All the wire types of well screens have higher intake-area ratios than any other present-day screen designs. Triangular-shaped wire screens have the lowest water intake velocity. The result is that less head or water pressure is lost when a triangular wire screen is employed than when any other type of screen is installed.

Other Screen Designs. Another major screen design is called the *louver* type. When viewed from the outside, the horizontal slots appear to be small louvers. The use of these screens is limited to gravel aquifers or gravel-paced wells.

Still another type of design consists of pipes that have been slotted vertically. The slotted screen can be manufactured, but more often it is made by torch cutting on the job site.

A continuous-wire screen will on the average have 8 to 10 times the open area that an equal-diameter slotted well screen has. Government tests made in Illinois demonstrated that a 10-foot length of 8-inch commercial, continuous wire screen performed four to six times better than a 20-foot length 12-inch, home-slotted pipe.

Plastic pipe performance or efficiency is greater than the home-slotted pipe because the plastic proportionally has more open area. Nevertheless, plastic pipe is only 1/6 to 1/10 as strong as stainless steel or Everdur. Therefore, plastic screens are limited to small-diameter wells that are not too deep.

Mechanical Considerations. Welded, continuous-slot screens are available in two series of diameters. One series is called *telescoping*. The other is called *pipe size*. The telescoping series of screens is designed to slip into the well pipe or casing. The outside diameter of these screens is a fraction less than the inside diameter of the pipe through which they are sized to pass. Typically, a 4-inch telescope-size screen actually has an outside diameter of only 3¾ inches. This gives the screen just enough clearance to pass through a standard 4-inch pipe and be lowered into place at the bottom of the casing or well pipe. Lowering the well screen through the casing is the method most often used for placing a screen because it is the surest and safest method. Telescoping screens are usually furnished with lead expansion rings at one end.

The series of well screens that are pipe-size, sometimes called ID screens, have inside diameters that are equal to the inside diameters of similar-sized standard pipe. The ID screens are used when it is necessary or desirable to maintain the same casing diameter all the way to the bottom of the well, or where for some other reason it is necessary that the well screen be permanently attached to the casing end and lowered with the casing. The ID well screens are manufactured with either welding rings at each end or pipe thread. The screens are designed to fit equal-size standard pipe. Nevertheless, threaded connections are rarely used with ID screens larger than 12 inches.

Slot openings for continuous-slot screens can be manufactured at practically any opening (from as little as 0.006 inch on up). The slot-opening designation refer directly to the opening in thousandths of an inch. For example, a screen with No. 10 slot size would have a spacing of 0.010 (or ten-thousandths of an inch). The size of the screen opening is crucial. Too large a slot opening will admit too much sand. Too small an opening and the flow of water will be greatly reduced.

Other Screens. Telescoping and ID screens are the two most often selected screens used for water well work. Another type is the louver or shutter type of screen. This too is manufactured in a variety of openings, generally in 5-foot lengths designed for welding in place.

Still another type is called the pipe-base screen. A heavily perforated steel pipe is used as its core. A continuous-slot screen tube is mounted over the pipe core. In some designs, the continuous wire screen is wrapped directly onto the surface of the pipe and welded in place. In another design, a number of bars or rods are spaced equally apart and welded to the pipe core. The continuous wire is then wrapped around the bars and welded to them.

Because the bars hold the wire a distance from the slots in the pipe core, the second type of pipe-core screen is more efficient than the first. Nevertheless, in both designs the water must first flow through the spaces that are in the screen and then through the pipe slots. The result is that there is much more pressure lost in the pipe-core screen than the straight screen design.

Still another variation of the pipe-core screen consists of welding a standard, prefabricated, continuous-wire screen onto a pierced pipe. It is supposedly still stronger than the two previously described pipe-core screens, but its efficiency is just as low. The purpose of the inner-core pipe is to increase the overall strength of the well screen.

Drive-Well Screens. Also called *drive points*, drive-well screens can be made by attaching a steel or bronze drive point to one end of a continuous-wire screen and a threaded collar to the other end. The drive point's maximum diameter is just where it screws fast to the screen. The drive point diameter is larger than that of the screen. This is done to help the drive point push stones past the sides of the screen and therefore protect it.

For further stone protection, a brass jacket well point is often used. This consists of a perforated bronze or brass pipe covered with a wire mesh screen that is in turn covered by a perforated brass sheet. This well point design carries screen protection just about as far as it can go. A variation of this eliminates the screen altogether. The core is a perforated or slotted brass pipe covered by a slotted brass tube spaced a distance from the inner pipe.

Whereas continuous-wire screen slot openings are specified by numbers directly related to the opening in thousandths of an inch, mesh-covered well-point openings are measured by the number of openings per lineal inch. Popular sizes run from 40, through 50, 60, 70, and 80.

ANALYZING THE AQUIFER

Sand and gravel aquifers that supply underground water the world over have been deposited there by either the actions of glaciers or streams. Detailed studies of these glacial and alluvial deposits have determined that the water-bearing materials are not positioned where they are now found by accident. The relative positions and grading of these stones, large and small, have been determined by various ancient geological activity.

The size or sizes of the grains of sand and the stones that comprise the aquifer into which you will dig or drill, and their mixture (if they are mixed) will determine the size of the openings of the well screen you should select. Mix or size distribution is an important consideration as well as the size of the individual grains of sand and the size of the stones. When all the grains or stones are alike in size, there is a physical limit to how tightly the sand or stones can be packed.

When there is a mix of sizes, packing can and generally is much tighter or closer. What happens is that the smaller particles of stone slip between the larger particles. For example, in a patch of gravel consisting of 1-inch roundish stones, no matter how tightly they were packed, there would still be a space between them large enough to encompass a stone 3/16 of an inch in diameter. If sand were mixed in with the 1-inch stones, the sand would fill the voids and the space between the stones would be reduced to the space between the grains of sand. See Fig. 11-2.

If the aquifer in which the screen is to be placed

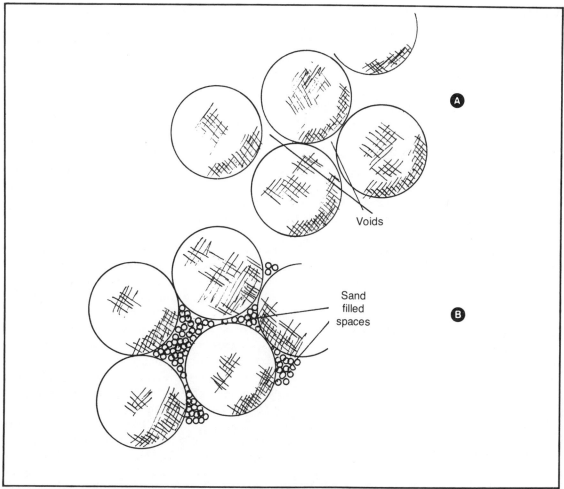

Fig. 11-2. When the gravel is comparatively large, as at A, fine sand will pack between the pieces of gravel and block the passage of water.

consisted of a mixture of fine and coarse sand, the permeability of the mixture would be far less than that of the coarse sand alone, but little more than that of the fine sand alone. This assumes that there is sufficient fine sand to fill the voids between the grains of coarse sand.

If the particles of stone comprising the sand and gravel mixture includes a range of sizes, or if the sand includes a range of sizes, permeability will not be significantly greater than that of the finest grains of sand contained in the mixture. In other words, the voids, the spaces between adjoining grains of sand, will not be larger than that which is

formed between the smallest particles of sand in the mix.

Analyzing Sand Samples. Professional well drillers plan and design their wells on the basis of formation analysis. They will first refer to whatever records are available to determine where their best chance of striking water at a reasonable depth exists. They will study the records to determine, at least roughly, what to expect in the way of overburden (topsoil), formation and the aquifer.

They will drive a test well in order to learn exactly what type of formation they will encounter and the composition of the aquifer as well as its depth and thickness. When they drill the test well, they will take constant samplings of the well core material removed. They will mark each sample carefully with the depth at which it was found.

This is impractical when you are driving a well point because you cannot see beforehand what the point is entering. When you jet a well, the flow of water brings the formation and the aquifer to the surface. Stone and sand will be mixed, but you will have a rough indication of the nature of the aquifer. All this is very useful in determining the best screen opening size.

Commercial well drillers always either analyze the sample material themselves or send it to a laboratory for analysis. Briefly, analysis consists of sifting a measured quantity of the samples through a series of sieves. Typically, they will use sieves with decreasing openings for analyzing sand and gravel (six screens for sand alone and five screens for fine sand).

The first and coarsest screen used is a 6-gauge

mesh with 0.131-inch openings. The last and finest has a 100-gauge mesh—with openings only 0.006 inch across. In use, the sand-gravel mixture is sifted through sieve one (the 6-gauge screen). Whatever passes through this sieve is sifted through the next smaller size sieve, and so on down through them all.

The material that is collected by each sieve is carefully weighed and the weight is recorded. The data so collected is used to draw a curve (Fig. 11-3).

Using the Data. The data on gravel and sand grain found by means of sampling and sieving is used to draw a graph. The graph is used to estimate the best screen- opening dimensions for a particular well. The most efficient openings in a well screen is that which at a minimum equals in percentage the voids or openings in the aquifer or the gravel pack, and at the same time can withstand the weight and pressure of the surrounding aquifer on the screen.

To explain, assume that the porosity or open areas of the surrounding aquifer is 25 percent. If the installed well screen has a porosity of no more than 15 percent, the water that flows from the aquifer through the well screen will be constructed by the well screen. There will be a greater pressure loss at the screen than in the adjacent formation.

In actual practice, the open area of the screen selected is often greater than the natural open area of the surrounding aquifer. The reason is that development will increase the porosity of the adjoining aquifer. Development removes the silt and fine sand from the aquifer alongside the screen, thereby increasing the porosity of the adjoining aquifer.

Selecting the Screen-Slot Openings. By rule of thumb, screen-slot opening selected range is from an opening that will *retain* 40 to 50 percent of the sand in the surrounding aquifer. This range of figures represents roughly the midrange of sand grain sizes. Put another way, a screen that will retain 50 percent of the sand will be too small for all the grains of sand larger than the grains that are halfway between the finest and the coarsest sand to be found in the aquifer. Put still another way, the 50-percent slot opening selected will stop grains larger than the opening and pass grains smaller than the opening.

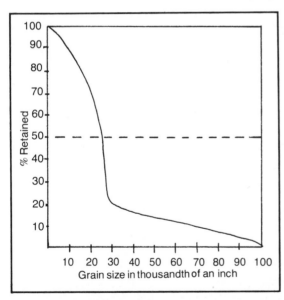

Fig. 11-3. Typical curve secured by passing sand (or gravel) through a series of screens of decreasingly smaller openings to determine the quantity of grains present in various sizes. The curve shows retention. If you wanted to keep sand grains larger than 30 thousandths of an inch out of your well, you would choose a screen with similar-sized openings. As you can see from the dotted line, 50 percent of the sand is finer than 30 thousandths of an inch.

While this might appear to be difficult, it is simple when the sand has been sieved and a graph drawn and the size mix is fairly even. This means that the line on the graph is fairly straight and the sizes are fairly equally divided. When this is the case, the vertical line on the graph, which represents percentage of retention, is followed to the right until it intersects with the grain-size curve. The point carried down to the required slot size is indicated by the grain size in thousandths of an inch.

Grains larger than the desired slot size, which are to the right of the size indicated, will not pass through. Grains smaller will pass through. The curve simply indicates how much of the sand grains are larger and how much of the grains are smaller than the percentage chosen.

When the curve is not smooth, it indicates that there is considerable abrupt variations in grain sizes. In such cases, a compromise slot-size opening is selected.

In some instances, it will be desirable to retain less than 50 percent of the surrounding sand. This is done when it is believed development will remove the fine sand from the immediate area around the well screen. Slightly larger slot openings are selected when the water quality is such that it tends to coat the slot walls with some sort of crust. To offset the effects of encrustation, larger slot-openings are selected. When the aquifer is immediately below a layer of very fine sand, slightly smaller openings are selected because, in time, the fine sand will be drawn down through the aquifer and into the well.

Gravel-Packed Wells. Wells are artificially gravel packed when it is believed that it will be easier and less time-consuming to pack the well artificially than it would be to develop the well naturally. Wells are artificially gravel packed when the aquifer is such that the well cannot be naturally gravel packed at all because the aquifer is fine sand and not a mixture of grain sizes. When there is a mixture, natural development will remove the fine and medium sand from alongside the screen. This leaves the coarse sand and gravel around the screen, which provides the packing.

Packing around a screen acts as a primary screen. This makes it possible to employ larger slots in the screen than might otherwise be practical. Larger screens means lower water restriction at the screen and greater yield.

Gravel size is determined on the basis of the sand that comprises the water-bearing formation. If the gravel is too coarse, fine sand can work its way into the voids in the gravel pack and so clog the

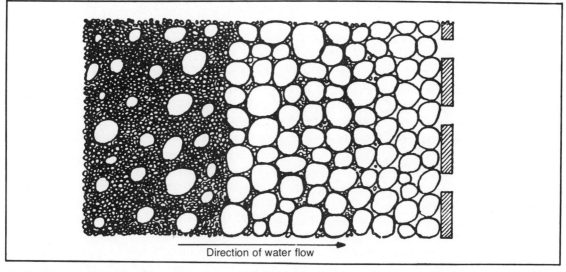

Direction of water flow

Fig. 11-4. Fine sand acts as a filter for the coarse sand (which in turn acts as a filter for the gravel). This shows the results of natural development.

gravel. If the gravel is too fine, the artificial pack could in itself reduce water passage. In addition to the care exercised in size selection, the gravel must be carefully screened so that all the stones are of the same size. See Fig. 11-4.

For medium-to-coarse sand formations, gravel up to ¼ inch in size is generally selected. For fine sand, gravel no more than ⅛ of an inch in diameter is used. Generally, gravel and slot size chosen is such that the screen will retain 75 to 90 percent of the gravel used.

Water Velocity. When water is pumped out of the well, the removed water is replaced by water from the aquifer that passes through the well screen and into the well. The speed or velocity of this water through the well screen has an important bearing upon the life and the efficiency of the well screen. Field and laboratory tests have proven that when this water travels at a velocity of 0.10 feet per second, friction losses through the screen slots will be at a minimum, encrustation rate will be minimal, and the rate of corrosion will be very low. Slower water speeds will be even more beneficial.

There is a formula by which water velocity through the screen openings can be calculated. For everyday, single-home, residential wells this is hardly necessary. To quote some relevant figures: a screen, 3 inches in diameter, 1 foot long with size 10 slots will pass 3.10 g/min; 40 slots will pass 9.9 g/min; and 60 slots will pass 13.0 g/min at a speed of 0.10 ft/min or less.

At first glance, 3 gallons a minute might not seem like much water, but that works out to 180 gallons an hour and 4320 gallons every 24 hours. That is a lot of water when the average household does not need more than several hundred gallons of water a day.

Screen Diameter. The prime consideration in selecting a well screen diameter is whether or not the well pump is going to be installed somewhere within the well casing. This is most often done with deep wells. An in-well turbine pump is selected because a pump near the bottom of the well is more efficient than one near the surface or on the surface. Only turbine pumps have the physical con-figuration (small diameter) at the required horse-power.

Well casings are not always the same diameter from top to bottom. On a deep well driven through a caving formation, it will be necessary to reduce the diameter of the casing with increasing depth. The pump diameter can then be selected on the basis of the casing diameter at the desired pump depth. The screen diameter can, in such instances, be less than the casing diameter. In any event, the pump's over-all diameter should be such as to provide a minimum of 1 inch of clearance between the pump and the well casing. By rule of thumb, the pump is selected two sizes less than the well-pipe diameter.

When working with a well point screen, the screen's diameter is close to that of the associated pipe. For very shallow driven wells, minimum-diameter pipe can be used. When you have to go below 20 feet and the formation offers increasing resistance to penetration, you have to switch to larger-diameter pipe just to provide the required stiffness. Screen diameter must be increased ac-cordingly.

When it is necessary to have a very long screen, which would only be the case in commercial installations, screen diameter cannot be minimum. The screen would be too weak. Although a screen should never be loaded with the weight of a string of pipe or casing, there is always the possibility that some loading will occur when the screen is low-ered. And a very small-diameter, long screen is difficult to handle without bending. When a long screen is required. the diameter is selected on the basis of minimal stiffness requirements.

Generally the last consideration in the selec-tion of screen diameter is yield. Doubling the diameter of the screen increases well yield about 10 percent. This rarely warrants the resultant in-crease in casing diameter that makes costs go up several times more than double. It also does not warrant the greatly increased costs of drilling the larger diameter hole.

Screen Length. Water well yield is directly related to well screen length. Double the length of the screen and you double its yield in gallons per minute. Obviously, you would like to have the greatest quantity of water available that is possible. At the same time, there is no need for yards of well screen to make available water you will never draw. Well screen length is therefore a compromise between your expected future needs, the yield per lineal foot of well screen, and the reduction in well efficiency that always accompanies the passage of time. With the years, some of the slots will become blocked up with sand. Mineral incrustation will reduce the openings of other slots. Corrosion will further reduce all the openings.

The first factor can only be accurately determined by drilling a test well and measuring the yield and extrapolating to find the minimum screen length required. This is what commercial drillers do in relatively unknown aquifers. An estimate can be made from data secured from neighboring wells (if there are any). If no information exists or is available and you are drilling blind, some indication of the potential yield can be drawn from the formation samples that are brought up.

Gravel aquifers generally provide the greatest yield. Fine sand is way down on the list and clay approaches the zero figure. If you are driving a point or jetting your well and it goes down fairly quickly, you can withdraw the screen and pipe and give it another try elsewhere. Do not jump to an immediate decision. Some well drillers report that they have been able to increase the yield of some of their wells nine times over by proper development work.

For maximum practical yield from a given aquifer of a limited thickness, well screen length must also be limited. By rule of thumb, when the aquifer is less than 25 feet thick, screen length should be held to within 70 percent of aquifer thickness. When the aquifer is thicker than 50 feet, screen length can be 80 percent of the aquifer's thickness.

In all aquifers it is usually good practice to position the screen well below the top of the aquifer. The reason being twofold. The deeper in the aquifer the less chance there is for fine sand seeping down into the aquifer as water is withdrawn. And if the water table drops, the lower the screen within the aquifer the better its chance of remaining supplied with water.

INSTALLING WELL SCREENS

When you drive a well point, the well point is the screen as well as the well point. With a well point, screen installation occurs simultaneously with the insertion of the point. When you jet a well and use a self-jetting well point, the same thing is true. Once these points are down to the desired depth, the well screen is in place. When other methods are used to construct the well, other screen-setting methods are used.

Care and Accuracy. The preceding methods of sinking a well can be termed self-screening methods. When neither of these two ways of positioning a screen are used, the screen is placed after the hole has been drilled to either its final depth or a short distance above its final depth. When this is done, considerable care must be exercised to keep accurate records of pipe length, cable length, and hole depth. You cannot "feel" the bottom of a hole when you have 50 feet or more of pipe and screen on your cable. Any mistake you may make here can cost you considerable time and materials.

Simplest Method. The simplest, most dependable and most used method of installing a screen is called the pull-back method. See Figs. 11-5, 11-6, and 11-7. The casing is sunk the full depth of the well hole. Whatever sand and debris present in the lower section of the casing is removed by bailing. The screen is then assembled. The bail plug is fastened to the bottom of the screen (which must be a telescoping type of screen). Then the lead packer fitting is screwed in place at the top of the screen.

The screen is lowered into place by means of the sand line, which is connected to a hook that engages the loop on the inside of the bail plug. Now comes the step that requires you to know exactly how much of the casing is in the ground. The casing is pulled back a foot or more short of the top of the screen; a portion of the screen remains within the casing.

The sand line is jiggled to unhook it and it is removed.

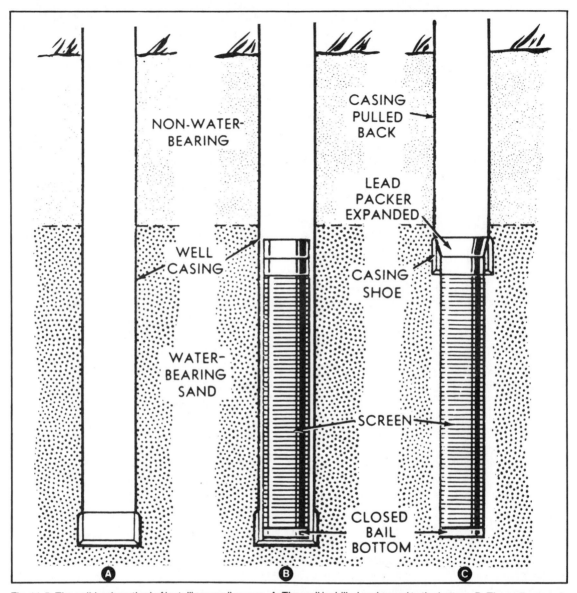

Fig. 11-5. The pull-back method of installing a well screen. A: The well is drilled and cased to the bottom. B: The well screen is inserted. C: The casing is pulled back. The lead packer is expanded and the screen is permanently exposed to the water-bearing stratum.

Next the lead packer is expanded to seal it and the screen inside the casing. This is done with a few light taps of a *swage block* or, as it is sometimes called, a *swedge*.

To lift the casing, you will have to put a casing ring or clamp on the casing pipe and position hydraulic jacks beneath the clamp to lift it.

Bail-Down Method. This method begins with drilling the hole down and setting the casing to a depth that will be a foot or more below the desired top of the screen when it is in position. Special fittings for the screen are required. One type consists of a bail-down shoe and guide pipe that are screwed fast to the bottom of the screen. A pipe is

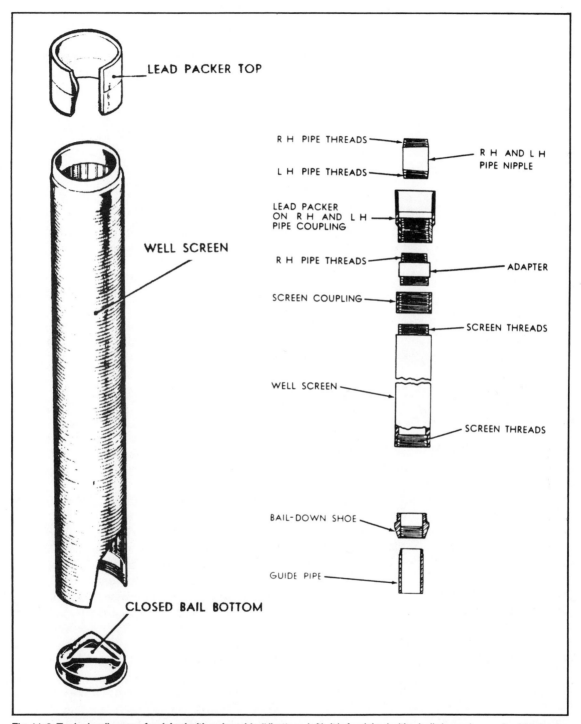

LEAD PACKER TOP

WELL SCREEN

CLOSED BAIL BOTTOM

R H PIPE THREADS

L H PIPE THREADS

R H AND L H PIPE NIPPLE

LEAD PACKER ON R H AND L H PIPE COUPLING

R H PIPE THREADS

ADAPTER

SCREEN COUPLING

SCREEN THREADS

WELL SCREEN

SCREEN THREADS

BAIL-DOWN SHOE

GUIDE PIPE

Fig. 11-6. Typical well screen furnished with a closed bail (bottom, left). It is furnished with a bail-down shoe and guide pipe (on the top left) and means for disconnecting a wash-down pipe once the screen has been positioned.

135

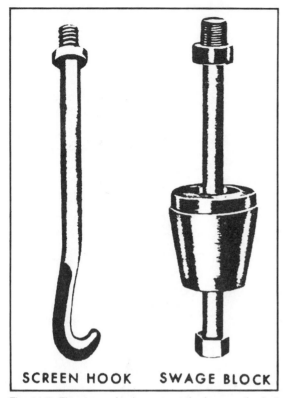

SCREEN HOOK   SWAGE BLOCK

Fig. 11-7. The screen hook engages the loop on the bail bottom. The swag block expands the lead packer.

then passed through the screen and connected to a special nipple with one left-hand and one right-hand thread. This nipple is screwed fast to the bail-down shoe. See Figs. 11-8 and 11-9.

The pipe or pipes are lowered through the casing until the shoe reaches the bottom of the casing. There it encounters the water-bearing sand. Now water is forced down through the pipe and out the shoe which includes a self-closing, one-way valve. The screen is now "jetted" down through the sand the desired distance.

The water is turned off. The sand outside the screen and the sand that has been driven up the annulus around the casing sink back, locking the screen in place. The pipe attached to the special nipple is next turned to the left, releasing it from the special nipple and permitting it to be removed. The value of this method is that no more than the final length of casing has to be used, and the last

distance at the bottom of the well does not have to be drilled.

Setting Pipe-Size (ID) Screen. When working in noncaving formations, the cost of the well is considerably reduced with pipe-size screen because a casing and an inner well pipe is not required. All that is needed is a string of pipe reaching from the screwed-on-screen to the surface. This pipe can be called a well pipe or a casing. In either case it is a single string of pipe.

In many instances, however, the bore hole is not noncaving from the surface of the ground all the way to the bottom of the hole. In such cases, the caving portion of the formation can be temporarily cased and the casing can be removed when the well is completed.

To position an ID screen and its attached pipe, the ID screen is screwed fast in place and then the entire assembly is carefully lowered into the well. Whether the hole is fully cased or partially cased, the ID must be very carefully lowered and supported by the cable until it is a foot or more from the bottom of the bore hole. Unless the well pipe is less than 50 feet or so, the weight of the pipe should never be permitted to rest on the screen. Pipe clamps of one type or another are then used to hold the ID screen and attached pipe in position until the surrounding soil moves in to lock the screen and pipe in position or the space around the pipe is grouted.

Rock Formations. Generally, when the water-bearing formation consists of consolidated or nearly consolidated rock, well screens are not required (Fig. 11-10). Water movement is through fissures and cracks in the rocks and no sand or silt is present. Temporary or permanent casing can be used to reach the rock formation. Once that has been reached, it will be penetrated a distance and then the diameter of the bore hole will be reduced a calculated amount. The drilling equipment—bit and cable—is then removed, and the permanent well pipe is lowered to the reduced-diameter bore hole in the rock. The pipe is now driven into the reduced-diameter hole. This locks the pipe in place and precludes the entrance of water and sand from the overburden.

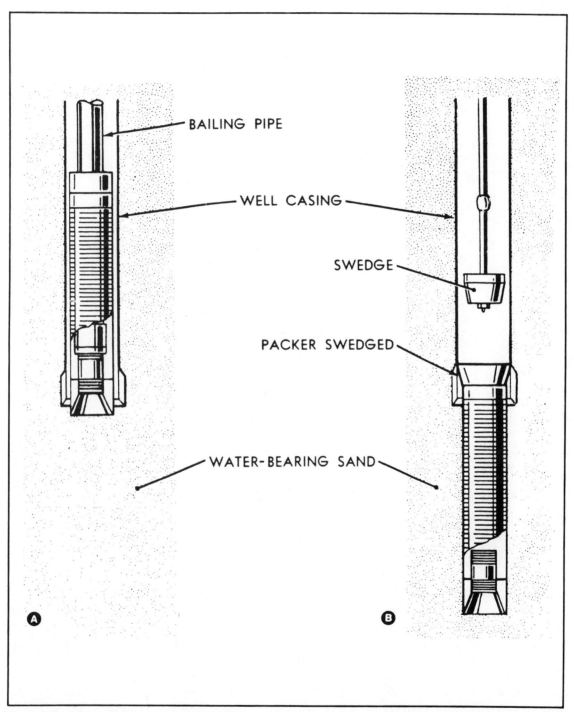

BAILING PIPE

WELL CASING

SWEDGE

PACKER SWEDGED

WATER-BEARING SAND

A

B

Fig. 11-8. Steps in the bail-down method of installing a well screen. A: Water pumped out the bottom of the bail-down fitting washes a path for the screen down through the sand. B: The casing has been pulled back and the swedge has been used to expand the lead packer.

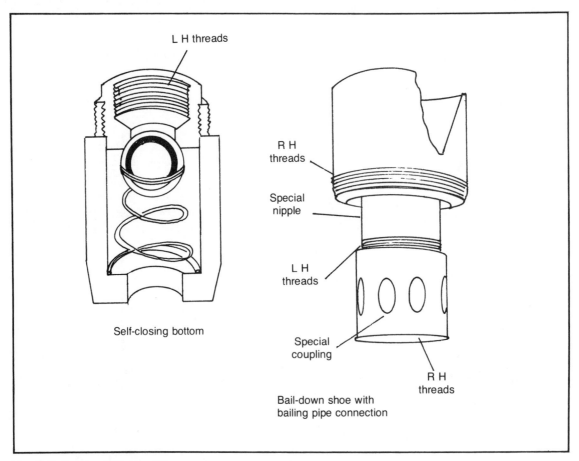

L H threads

R H
threads

Special
nipple

L H
threads

Special
coupling

R H
threads

Self-closing bottom

Bail-down shoe with
bailing pipe connection

Fig. 11-9. Two special bottom fittings.

RECOVERING AN IN-PLACE SCREEN

One of the advantages of using a telescoping well screen is that it can be removed from the well without disturbing the well pipe or the well casing. This cannot be done with ID screens.

Screens are removed from wells when it has been found that the slot openings are entirely wrong for the aquifer in which they have been placed. Screens are sometimes removed when a well has to be abandoned and the screen can be used elsewhere. Screens are removed when they have become blocked or badly corroded.

Sometimes a screen will become so badly covered with scale or incrustation that it seriously reduces water yield. When any of the above occurs, the screen has to be removed. There are certain chemicals that will dissolve some forms of scale and incrustation. Nevertheless, it is generally not recommended that they be used. It is safer to remove the screen by the sand joint method.

The Sand Joint Method. Essentially, it consists of constructing a sand joint, attaching it to the end of a string of pipes, lowering the joint to the bottom of the screen, and then pulling the joint and the screen up and out of the well casing.

As you can imagine, considerable force is required to do this. This is so much that any old pipe cannot be used to pull the sand joint and screen up and out. For a screen 3 inches in diameter, it is recommended that a 1-inch pipe be used. For a 3½-inch pipe, a 1¼-inch pipe should be used. For a 4-inch screen, a 1½-inch pipe should be used. For a

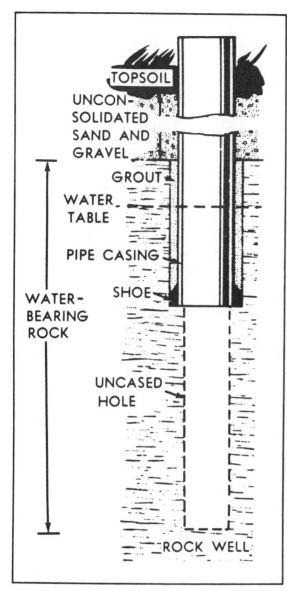

TOPSOIL

UNCON-
SOLIDATED
SAND AND
GRAVEL

GROUT

WATER
TABLE

PIPE CASING

SHOE

WATER-
BEARING
ROCK

UNCASED
HOLE

ROCK WELL

Fig. 11-10. When the casing can be forced into a watertight fitting in a rock formation, a well screen is sometimes omitted.

5-inch screen, 2½-inch pulling pipe should be used.

The force required amounts to many tons. Upward pressure can be provided by a drilling rig or a pair of heavy-duty hydraulic jacks and a strong pipe clamp (or perhaps two, one above the other to prevent slippage).

Constructing the Sand Joint. Figure 11-11 shows the assembled sand joint at the bottom of the well. The sand joint is constructed by wiring strips of sacking to the pulling pipe just above a screwed-in-place pipe cap. The sacking strips are cut 2 to 4 inches wide, depending upon the diameter of the well screen. The length of the pieces is such that after one end of a strip has been wired to the pipe, there is sufficient coarse cloth to reach the inside of the screen and up to its top. The strips are dispersed evenly around the pulling pipe.

The sacking, which may be burlap, is lowered a short distance into the casing. The strips should form a rough bag with their edges overlapping. A little sharp, clean sand is now placed in the "bag." The sand keeps the strips up against the casing wall. The pulling pipe and the bag are lowered a little more. Additional sand is added. The bag is lowered some more and more sand is added. This is done until the full length of the bag is inside the casing. The cleanliness of the sand is very important. If any clay is mixed in, the joint will not lock in place.

Using the Joint. The joint is slowly lowered to the bottom of the well where it will rest on the bottom of the well screen. The theory behind this joint is that, as long as the joint is immobile or moving downward, the sand that fills the bag is loosely held up against the bag and the adjoining casing wall.

When the pulling pipe is slowly and carefully pulled upward, the bottom of the bag tends to expand as the pipe cap presses upward against the bag filled with sand. Because sand is almost incompressible, the upward pressure of the bag and cap against the bottom of the sand tends to push the sand outward. This results in locking the sand and its bag to the inside of the screen. Wet sand works better than dry sand. Once the bag and pipe lock against the casing, the pipe can be pulled up—bringing the screen up with it.

Should there be a need to remove the sand joint without displacing the screen, this can be done by washing the sand away. This can be accomplished with a jet of water or a jet of compressed air.

Acid Treatment. Treating the screen of an old well with acid sometimes helps loosen the

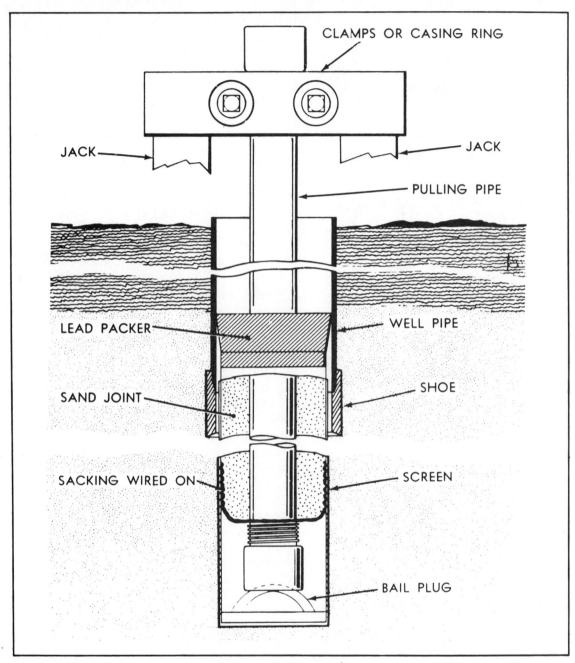

Fig. 11-11. Details of a sand joint.

screen and reduces the lifting force needed to break the screen loose. The screen is filled with a mixture of muriatic acid and water at a ratio of 1:1 by volume. Plastic or rubber hose is used to lead the acid directly into the screen. The acid is left in place overnight or at least several hours. Then the acid and water mix is pumped out before the sand joint is employed.

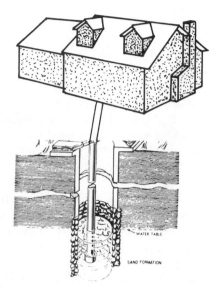

Chapter 12
Packing
and Developing

P ACKING IS THE TERM USED TO DESCRIBE THE placing of a layer of gravel just outside the well screen. The layer of gravel, the packing, is usually several inches thick and extends completely around the well screen and somewhat higher than the well screen. When packing is properly positioned, water from the aquifer must first pass through the packing to reach the well screen.

The purpose of the packing is to provide a primary, coarse screen in the path of water flow preceding the passage of the water through the well screen. The use of packing is arbitrary and depends upon the judgment of the well driller. Packing is generally not used in gravel and coarse sand aquifers. Generally, it is used when the aquifer consists or contains a high percentage of fine sand. The gravel packing stops the fine sand and permits the use of larger screen slot openings than would otherwise be safe to use. Larger slot openings (without the entrance of sand) increases the yield of the well.

Gravel Used. The gravel used for packing is on the order of ⅛-inch across and smaller. In the absence of a firm basis for the selection of any particular size gravel, coarse sand is often employed.

It should be noted that the size of gravel larger than coarse sand must be selected with great care. If the gravel is too large for the sand it is to retain, the sand will fill the voids in the gravel and prevent the passage of water. That, of course, is exactly what is not the object of installing packing.

The gravel used must be carefully screened for size so that all the stones are very close in size: no stones larger and no stones smaller than what is desired. Prepared gravel can be purchased from well-drilling equipment supply houses. Their addresses can be found in the phone book or Thomas (a commercial listing of all companies. You will find these books in the library). To use any old gravel consisting of a range of sizes is worse than useless. When gravel consists of many different size pieces, the gravel tends to pack and become relatively impervious to the passage of water.

Need for Packing. The purpose of gravel packing is to provide a primary screen to intercept the movement of fine sand to the well screen. There is no arguing that statement. Gravel packing works. Keep in mind, however, that proper well development can also produce the equivalent of gravel packing.

In theory this is possible in all wells. In practice it is not so. Two factors are involved. One is the nature of the aquifer. Thorough well development will remove the fine sand from around the well screen, leaving coarse sand and gravel. That is exactly the same thing as if the well was artificially packed. In some aquifers this is difficult to do and in some aquifers it is nearly impossible to accomplish (for example, if the aquifer is nothing but fine sand).

The other major factor involved in deciding whether or not to pack the well artificially is cost. Gravel packing always requires additional material and additional labor. The quantities of each depends upon the type of well drilled, its diameter, and the method used to place the gravel.

Both of these considerations must be weighed against how much time and effort is involved in developing a natural gravel pack. The time required to develop a well naturally can only be estimated on the basis of experience.

The Importance of Data. So now comes the experience factor or however you might want to define the knowledge needed to balance the nature of the aquifer and the time required to develop that aquifer naturally against the costs of artificial development. When a decision has been made to artificially gravel pack the well and gravel size has been selected, any of the following methods can be used, provided they are suited to the well.

GRAVEL PACKING METHODS

Pull-Back. Often called the *positive method* (Fig. 12-1) because it never fails, the *pull-back method* is simple in execution but expensive in materials. The well must be *cased*. That means there is a well pipe extending from the well screen up through an outer casing. Relative casing and well-pipe diameters must be such that there is approximately 2 inches of annular space between them. The packing will fit into this space. Therefore the thickness of the packing layer will be determined by this space.

While packing thickness is not crucial, it is important. Too thin and the packing is ineffective. Too thick and it would be nearly impossible to remove mud present on the walls of the well. This is a problem with rotary drilling that depends upon mud to prevent the bore hole walls from caving. The mud is pressed against the wall and forms an almost watertight barrier. When the artificial packing is more than several inches thick, it is difficult to remove the mud layer.

The pull-back method is simple. The well is drilled. The casing is driven to the very bottom. The well pipe with the screen attached is positioned in its place at the bottom of the bore hole. To keep the screen centered, a spring guide is fastened to its bottom. Next, the gravel is gently poured down the annulus. Water can be added to aid the flow of gravel. Sufficient gravel is added to bring the top of the gravel pack a few feet above the top of the screen. To estimate this distance, the gravel is deposited in measured quantities after the annual clearance has been calculated. The casing is now pulled back sufficiently to expose the screen and the gravel around it.

If you have drilled into a caving formation and there was no alternative, the well casing is already in place and little is wasted beyond the discarded (or reused) upper section of casing.

Bail-Down. This method may be used when the well is sunk by the bail-down method. See Fig. 12-2. A bailing shoe somewhat larger than the screen is used. As the screen is bailed down into the sand aquifer, gravel and coarse sand is placed within the annulus. The packing material follows the screen down.

Noncaving Formations. When the lower reaches of the bore hole do not tend to cave-in and can be held in place by relatively clean water alone, the gravel can be positioned by means of a termie. This is a long pipe with a funnel at its top end. Its bottom is positioned in the annulus. Gravel poured into the funnel is guided directly into position. This prevents the gravel from bridging and jamming up somewhere along its journey downward, and the termie also prevents the falling stones from knocking down rocks and dirt during its passage. The goal

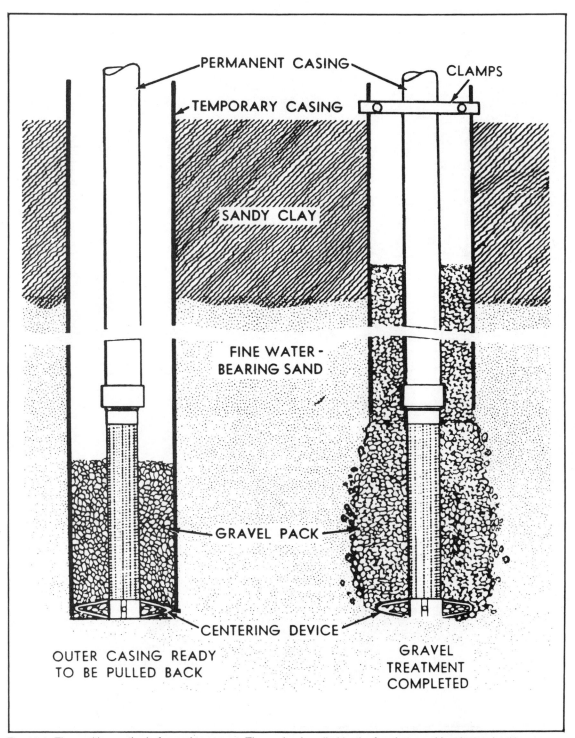

PERMANENT CASING

CLAMPS

TEMPORARY CASING

SANDY CLAY

FINE WATER-
BEARING SAND

GRAVEL PACK

CENTERING DEVICE

OUTER CASING READY
TO BE PULLED BACK

GRAVEL
TREATMENT
COMPLETED

Fig. 12-1. The positive method of gravel treatment. The casing is pulled back after the gravel has been placed.

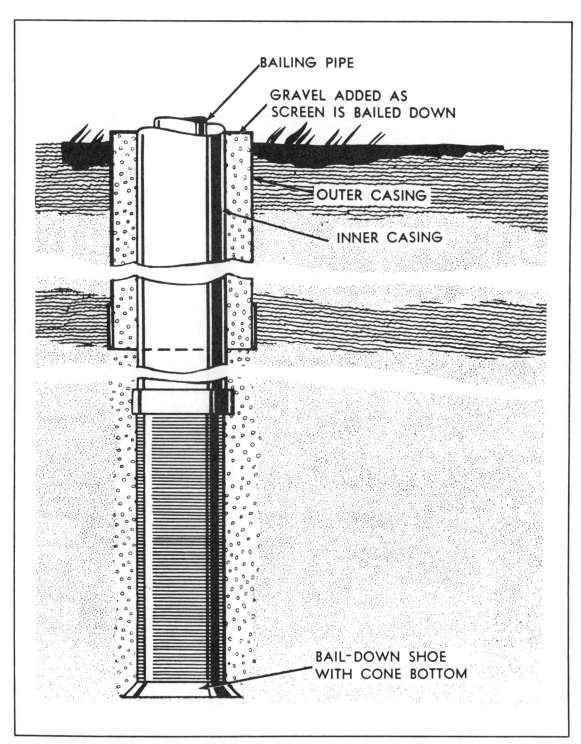

BAILING PIPE

GRAVEL ADDED AS
SCREEN IS BAILED DOWN

OUTER CASING

INNER CASING

BAIL-DOWN SHOE
WITH CONE BOTTOM

Fig. 12-2. Packing gravel can be dropped through the annulus when a screen is washed down into place.

144

is to have a one-size blanket of gravel around the screen. Any unwanted addition reduces its effectiveness.

Slot Size. When the well is artificially packed, it is important to select a slot opening that retains 75 to 90 percent of the gravel. This means that the slot openings will be comparatively large. It will be much larger than the slot openings chosen when the goal is to keep out the fine sand.

Fine-Sand Alternative. When the aquifer is composed only of fine sand, natural development will not work, but artificial packing is only one answer to the problem. Another solution is to use a screen with fine slots and to increase the length of the screen proportionally. This might not be practical on a commercial installation where the screen might already run to 20 or 30 feet. It is a practical solution for a residential well driven or jetted without a casing.

DEVELOPMENT

The purpose of "natural" development is to create a gravel pack immediately outside and around the well screen. Constructing a well by any means always disturbs the aquifer. With some methods the disturbance is moderate. With rotary drilling employing mud, not only is the aquifer disturbed, but the mud is driven into the voids in the aquifer and often ends up as a thick crust or cake on the surface of the bore hole. The overall result is that whatever water might have moved from the aquifer into the well is considerably reduced. Goal number one of development is to return the aquifer to its original condition by removing the barriers produced by drilling.

Goal number two is to remove any fine sand and silt present in the immediate area around the well screen. This has to be done or the well will be a *sand pumper;* sand and silt will find its way into the water. Otherwise screen openings so fine have to be used that the well's yield is reduced to an unacceptable figure. See Figs. 12-3 and 12-4.

Theory. In the normal course of events, simply pumping the well will not bring about the desired results. True, in time the worst well probably will run clear. The action of the pump drawing

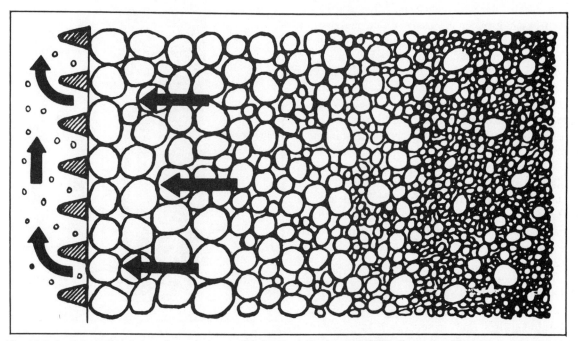

Fig. 12-3. A cross-sectional view of a perfect example of natural development. All the fine grains of sand have been drawn through the coarser spaces in the gravel and into the well from where the sand is removed by bailing.

145

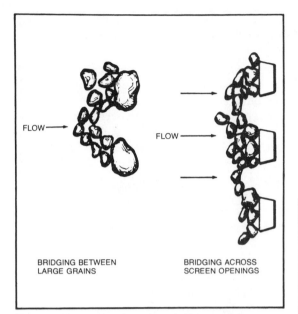

Fig. 12-4. How grains of sand can bridge openings.

water out of the aquifer around the screen will eventually pull all the fine sand and silt into the well.

The trouble with that can be summed up by the term *bridging*. When the grains of sand move in one direction only, they tend to clump up and form "bridges" across the slot openings and across the voids between pieces of gravel. Simply drawing water out of a well can clear the water, but it will almost always result in decreased yield. What is required is a vigorous in and out flow of water; a strong, cyclic flow reversal for as long as it takes to produce the natural packing that is needed.

Surge Plungers. A surge plunger is a kind of piston that is moved up and down in the well pipe. It can be made on the job or purchased. There are two types. One has a valve on the bottom that reduces the effectiveness of the piston action. That is what you want at the beginning of development. See Fig. 12-5.

The plunger is weighted to keep the cable taut. The plunger is then lowered into the well until it is a few feet above the top of the screen. The plunger is moved slowly at first, raising it 3 to 4 feet and then lowering it. This is done for several minutes. Then remove the plunger and, with a bail, remove what-

ever sand has been drawn into the well. Go back to working the plunger again. This time it can be operated a little more rapidly.

After a while, stop the surging and remove the sand. Keep track of the amounts of sand removed each time to keep track of the development progress. Increase the surging time as the quantity of sand removed from the well decreases. Stop surging and bailing when no more sand is brought into the well.

This process can take as little as a few hours or

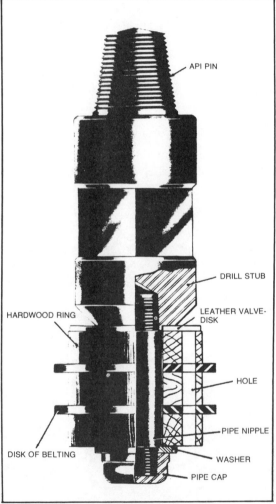

Fig. 12-5. A commercial surge block. The same action can be duplicated with a simple block-and-leather disks arrangement.

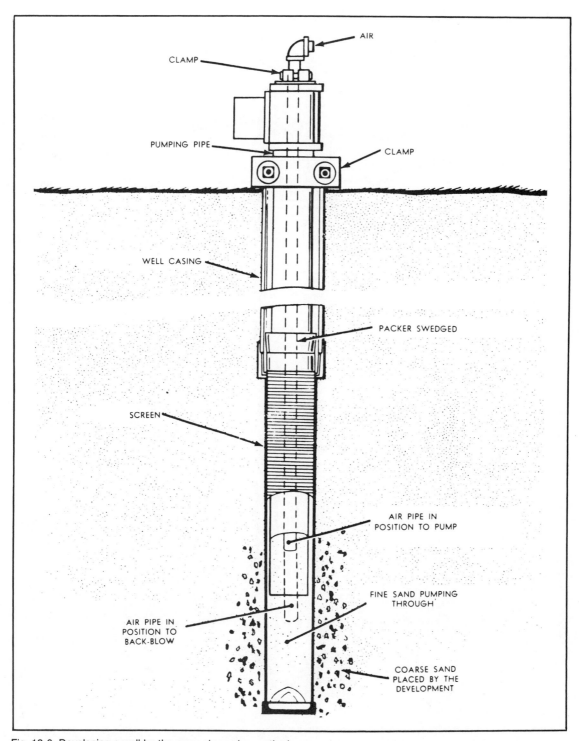

Fig. 12-6. Developing a well by the open-air surging method.

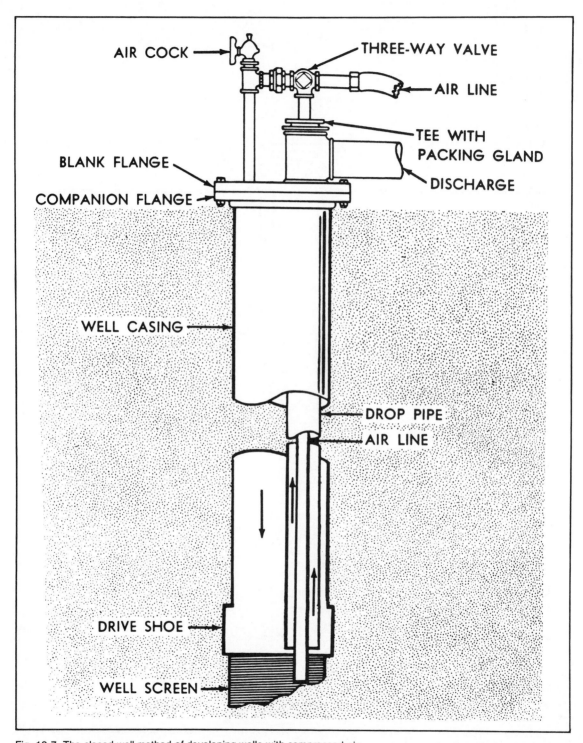

AIR COCK

THREE-WAY VALVE

AIR LINE

TEE WITH
PACKING GLAND

BLANK FLANGE

DISCHARGE

COMPANION FLANGE

WELL CASING

DROP PIPE

AIR LINE

DRIVE SHOE

WELL SCREEN

Fig. 12-7. The closed-well method of developing wells with compressed air.

as much as three or four days. It all depends on the size of the well and the nature of the aquifer.

Developing With Compressed Air. The water is driven in and out of the well by means of blasts of compressed air. A pressure of about 100 psi or higher is required, along with a capacity of about 250 to 600 cfm/min of air. Depending upon the depth of the well and the quantity of water above the end of the air pipe, some 100 to 150 gpm can be pumped out of the well by air lift. In addition, a drop pipe, a compressed air tank, and a quick-acting, two-way air valve are required along with some flexible air lines and piping. See Fig. 12-6.

When a 4-inch or larger well is to be developed, with air, the drop pipe should be 2½ inches. The air line should be ¾ of an inch or larger.

The drop pipe is positioned within the well with its lower end about 2 feet from the bottom of the screen. The air hose or pipe is placed within 1 foot of the bottom. Sacking is wired around the top of the air pipe, where it enters the well to keep water from shooting out here.

The first step consists of starting the compressor and turning the valve to direct air into the air-storage tank. When this is up to maximum pressure, the valve is turned to direct air down the air pipe. The air emerging from the bottom of the air pipe pumps water and sand out of the well. When the water runs fairly clear, the valve is turned to direct the air in the tank down the air pipe. This produces a blast of air that drives the water inside the screen back out and into the formation.

This method will work only when there is a considerable head of water above the end of the air pipe. With only a little water, the blast of air will simply push the well water up the casing. The storage tank is again filled with air, closed off, and the air is used to again pump the well. This is repeated as often as necessary to move both the drop pipe and the air pipe up a little each time to expose a different portion of the screen to the air blast. Another method of developing a well by means of compressed air depends upon pressure being built up inside the well to drive the water out through the screen. See Fig. 12-7. This method requires a compressor, drop pipe and airline, plus a

means of temporarily sealing the top of the casing, an air cock, and a three-way valve.

In operation, the drop pipe and the air line are lowered into the well all the way down into the screen. Compressed air is introduced into the air line and this drives the water and accompanying sand up and out of the well. The air cock is closed as is the water discharge pipe. The three-way cock is used to direct air down the well casing.

As air pressure builds up within the casing, the water at the bottom is pushed out through the screen. The air supply is shut off, the air cock is opened, and the water discharge pipe is opened. The underground water now refills the well.

When air stops hissing out of the air cock, the well is as full of water as it will ever be. The process is then repeated again and again until air pumping brings up nothing but clear water.

Development by Backwashing. Because any cyclic motion of water in and out of the well screen will produce a natural gravel pack around the screen, any arrangement that does this will produce the desired results. See Fig. 12-8.

One technique consists of rapidly pumping the well dry or almost dry and then letting the pumped water flow back down in a rush. Another method consists of using a jet of water to produce the desired backwash. Still another technique consists of building water pressure within the well and then suddenly releasing the pressure.

Developing Rock Formations. All the previous methods described for well development can be used in rock formations. It is customary to increase the size and number of cracks and other openings through which water enters. An acid that dissolves limestone is used.

Redeveloping Old Wells. With the passage of time, well yield drops due to incrustation and scale deposits if that is the nature of the water. If the water is acidic, the well screen will be eaten away. Incrustation and scale deposits are self-growth phenomena. When the slots first decrease in size, the passage of water slows. When the flow of water slows, more minerals are deposited. This resudes the slot areas and further flows the flow of water. In time the slots can be completely closed.

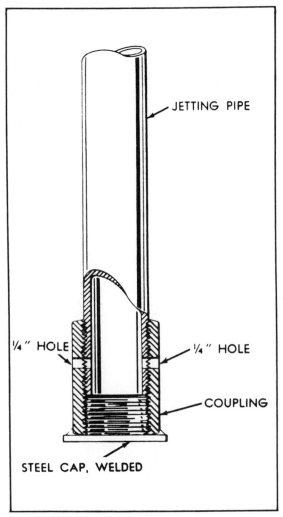

JETTING PIPE

¼″ HOLE

¼″ HOLE

COUPLING

STEEL CAP, WELDED

Fig. 12-8. Improvised jetting tool.

Any of the aforementioned well development methods can be used to increase the yield of an old well. If they are used early enough, the scale and incrustation can be loosened and washed away. If these methods do not work, then an acid or chemical treatment must be used.

Sufficient muriatic acid is released by means of a pipe to fill the area behind the screen. The acid is permitted to remain in place for an hour or two, and then the screen is lightly surged. Some two to four hours more are then permitted to pass. The well is then bailed or pumped free of the acid and surged some more. Following, the well is bailed free of whatever debris the acid and surging has brought into the screen.

If the acid treatment does not produce results, try a solution of hexametaphosphate. Mix sufficient amount of the chemical into a 50-gallon barrel to produce a solution of 3000 to 5000 ppm (parts per million). Put another way, use 4 pounds of the chemical to 100 gallons of water standing in the well. In addition, add calcium hypochlorite to produce a solution of 50 to 100 ppm of the chlorine. Permit both solutions to remain in the well for an hour, then surge it vigorously for 3 hours, and pump the solution and debris out. Repeat this procedure three or four times.

Alternative. If none of the above treatments increase the yield satisfactorily, there is no alternative to removing the screen and cleaning it by hand or replacing the screen. If an old well starts to pump sand, the screen has corroded through and must be replaced.

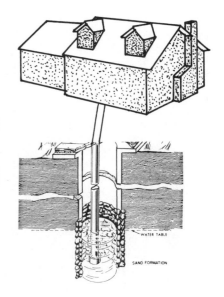

Chapter 13

Well Completion

A PRIME CONCERN WHEN SELECTING THE SITE of a well is the possibility of pollution. That is why all possible sources of contamination must be carefully considered before well construction is even begun. Proper siting is not enough. The well must also be protected from surface water contamination that either drains down the annular space around the well pipe or casing and reaches the well screen or leaches through the earth and then moves sideways to reach the casing and travel downward.

GROUTING

Grouting is the time-tested method of sealing the annular space around the well pipe or the well casing with cement positioned under pressure. Pressure grouting not only seals the space between the hole and well pipe or liner, it also seals the cracks and crevices in the formation and protects the well casing and or pipe from corrosion. Some commercial well drillers report that properly pressure-grouted casings have withstood corrosion for 30 or more years. Pressure grouting is also used

to seal off undesirable aquifers in situations where it is necessary to drill through one aquifer to reach another because the first (upper) aquifer is contaminated or is carrying unpotable water.

Pressure cement grouting, when properly accomplished, limits the water that enters a well screen to the water present in the aquifer in which the screen is immersed. No other water enters. To be certain of this, the grout must extend several feet below the lowest depth to which the well will be pumped. See Fig. 13-1.

Techniques. The technique or method selected will depend upon the depth of the well and the nature and size of the bore hole. No particular method is best for all wells and the well driller has to improvise and adjust to the individual condition.

A driven well cannot be grouted because there is no open space between the riser or well pipe and the surrounding formation. In such cases, the upper few feet of pipe can be exposed by hand digging and that open space filled with cement. Here you can use a mix of 2 parts sand to 1 part portland cement.

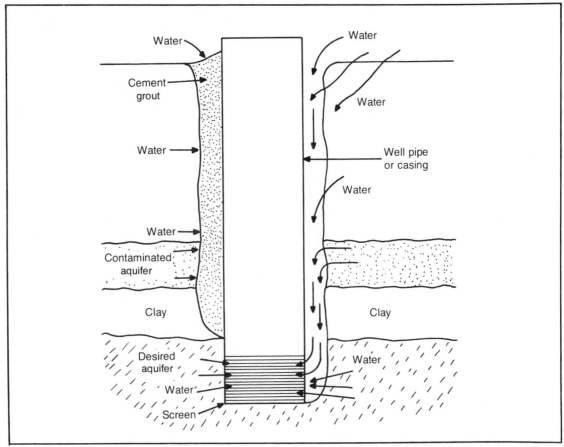

Fig. 13-1. The left side of the well in this diagram has been properly sealed with cement grout. Neither surface water nor water from the contaminated aquifer can get into the well. The right side has not been sealed. Surface water and contaminated water can enter the well.

The same would be true of a self-jetting well.

When the bore hole is self-supporting, at least for a distance below the surface all the way down to the well screen, the screen must be protected from the grout. Otherwise, should the grout surround the screen, no water will enter the well. The cement can be prevented from entering the annulus around the screen by filling this space with graded gravel of the proper size in relation to the screen slot openings.

In effect, this is artificial packing. If the screen has already been packed, the packing should be topped by an additional foot or more of gravel. This can be gravel of any size or mix because you do not particularly want water coming down into the

gravel pack from above the screen. Other means can also be used to seal the screen against the entrance of liquid cement.

If the screen has been exposed by the draw-back method, but has not been packed, give the aquifer time enough to consolidate around the screen. To speed the process, this well could be pumped. To be certain no cement works its way into the screen, several vertical feet of gravel can be placed within the casing atop the lead packer. If the screen is positioned deeply within the aquifer, you do not need these precautions because wet cement will not go through more than a foot of small gravel, and even less of coarse sand.

To grout a well, use a mixture of portland

cement and water on the order of 1 bag of cement to 5½ to 6 gallons of water. To secure a more free-flowing mix, add about 2 to 6 percent finely ground bentonite to the mixture of cement and water. Bentonite is a special clay that is often added to drilling mud. The bentonite, in addition to making the mix more fluid, will reduce shrinkage (which is normal to all portland cement).

The cement has to be thoroughly mixed and run through a screen having a ¼-inch hole, to make certain no lumps are in the mix. The cement can be hand mixed or power mixed or purchased from a concrete company and hauled to the job site in a truck. Quantity can be estimated by calculating the open annular space and adding a percentage for waste and error.

Above ground, on a warm day, a portland/water mix will set in about 1 hour, harden permanently in about 36 hours, and reach 75 percent of its ultimate strength in about one month. In the

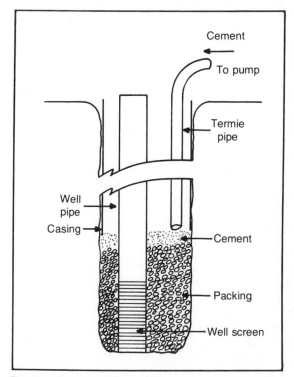

Fig. 13-2. A termie pipe is used to pump grout into the wall's annulus. Packing around the screen keeps the liquid cement from sealing the screen.

ground, where the temperature might be in the 60s setting (stiffening) could take several hours.

Once the cement has been mixed with water, the chemical process that takes place is irreversible. It is only a matter of time before the mixture turns rock hard. If you place the still-wet mixture in a gunny sack and toss the sack into a lake, the concrete or cement will harden to the shape of the sack. If you sprinkle the same mixture directly into the lake, you will end up with a lot of fine, small concrete or cement stones. You cannot simply pour the mix down anything but a very shallow annulus and secure the solid stone barrier you need.

The grout is positioned by means of a pipe. In this case, it should be no smaller than ¾ of an inch, with a larger diameter pipe preferable. See Figs. 13-2 and 13-3. One end of the termie is positioned just above the beginning of the packing. The other end of the termie is connected to a pump capable of driving the liquid grout down the termie at a pressure of 100 psi or whatever is considerably greater than the head (pressure) of the water at the bottom of the well.

Simply pouring the cement down the well will result in sand. The cement has to be deposited under pressure so that the water is moved up. Because the grout should extend below the well's water level when it pumped at maximum, the end of the termie will always be under water. The termie is raised as the annulus is filled. Pumping is stopped when the cement runs out of the annulus and onto the ground. See Fig. 13-4.

DISINFECTION

In the course of constructing any well, the piping and the tools used cannot fail but to pick up some soil. Soil almost always contains pathogens (disease causing organisms). To disinfect the well before use, the well and associate piping and pump should be disinfected.

The simplest method is to use a chlorine solution. Generally sodium hypochlorite is used. This is prepared by bubbling chlorine gas through a solution of caustic soda. Common household bleach contains 5 percent available chlorine. To sterilize a well, a solution containing about 100 ppm of avail-

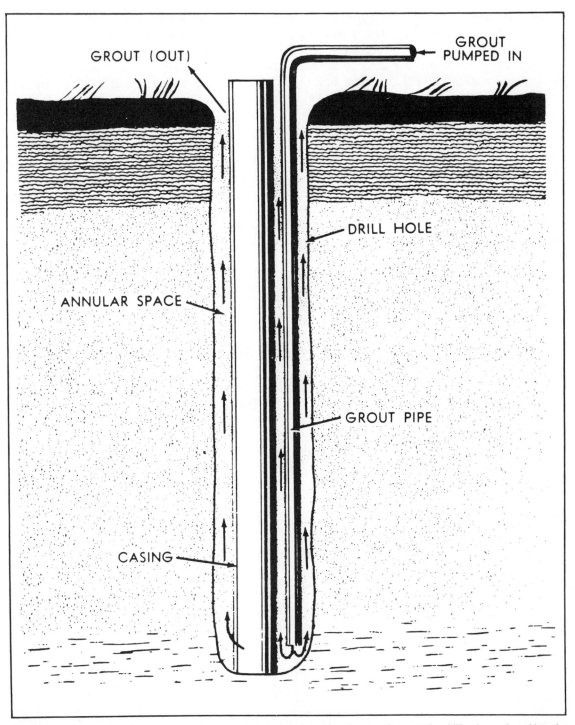

GROUT (OUT)

GROUT PUMPED IN

DRILL HOLE

ANNULAR SPACE

GROUT PIPE

CASING

Fig. 13-3. In this arrangement, the well is grouted before completion. After the grout has set, the drilling is continued into the aquifer.

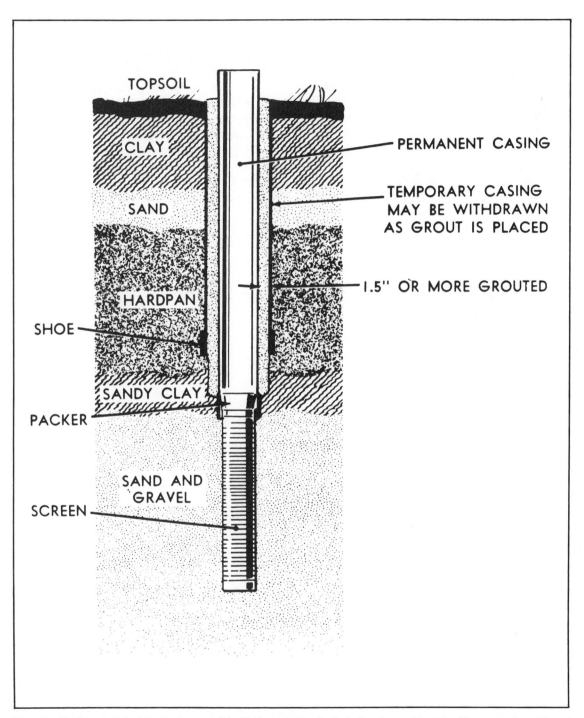

TOPSOIL

CLAY

SAND

HARDPAN

SHOE

SANDY CLAY

PACKER

SAND AND GRAVEL

SCREEN

PERMANENT CASING

TEMPORARY CASING
MAY BE WITHDRAWN
AS GROUT IS PLACED

1.5" OR MORE GROUTED

Fig. 13-4. The bore hole is drilled large enough for the inner casing (well pipe) and an outer casing. The screen is positioned and the packer is expanded. The space between the well pipe and the outer casing is grouted. As it is grouted, the outer casing is slowly removed.

able chlorine should be used. To secure this concentration, it is necessary to mix 0.4 quarts of 5 percent chlorine bleach to every 100 gallons of water in the well and associate piping and pump.

The chlorine must be agitated in the well and pumping equipment and should be permitted to remain in place for four hours or more. It is also necessary to disinfect all gravel-pack material. It is a good practice to pour a little of the chlorine into the well while the drilling tools are in place. If the well is very deep, it will be difficult to achieve the required chlorine concentration with a liquid. In such cases, dry calcium hypochlorite is placed in a suitable container and lowered to the bottom of the well. The container is moved up and down until the chemical has been dissolved.

To make certain the entire water system is being disinfected, the chlorinated well water should be pumped through the plumbing system and all the faucets opened until the smell of chlorine is obvious. The faucets are then closed and the chlorine is permitted to do its work. Water is then pumped through the system until the smell disappears and the water no longer feels soapy.

Part 2

Septic Tanks
and
Drain Fields

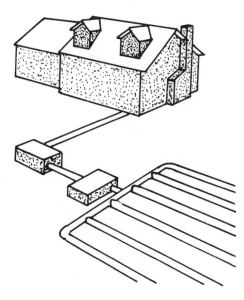

Chapter 14

Septic Systems and Ecology

THIS PART OF THIS BOOK COVERS THE OPER-
ation, design, construction, and service of
septic tanks and drain fields. The material is presented in the order, more or less, in which the
sewage leaves the building and travels through the
septic tank to the drain fields and then into the
earth. This is convenient from both the writing
point of view and the readers point of view.

If you are going to use this book as a practical
guide to constructing and installing a septic tank and
drain field, the preceding description is *not* the
correct way to use this book. In actual practice, that
is not the correct sequence. The correct sequence
begins with a *percolation test* (see Chapter 21). First
you must determine whether or not the soil on your
property will absorb water at a satisfactory rate. If
it will there is no problem. If it will not absorb water
satisfactorily, study the methods suggested for
making soil more permeable; increasing its rate of
absorption. If this cannot be accomplished in a
practical manner, then you *cannot* install a drain
field. Without a drain field or a *leaching field* as it is

sometimes called, there is no point in attempting to
install a septic tank; the tank cannot drain.

Read this entire book, but check the section on
percolation carefully before you begin any construction work.

The first portion of this book describes the
drilling of water wells. Land outside a municipal
water system is generally considerably less expensive than land within reach of a system. By providing your own water, far less costly property can be
used for building your home.

The prime reason for constructing a septic
tank system is also economic. Doing so makes less
costly real estate useful as a home or farm site. In
many instances, you will find building lots with city
water, but without city sewers. The reason is that it
is far less costly to install a water line than it is to
install a sewer line.

ECOLOGICAL CONSIDERATIONS

There is another reason for installing a septic tank
system. It helps our ecology and our economy.

During a normal year, the average American produces some 3.5 pounds of phosphorus and 9.9 pounds of nitrogen as body waste. Treating this waste to safely dispose of it at a central, municipal plant is expensive in equipment and energy. These plants use petroleum, methane, lime carbon, alum and lime, or ferric iron compounds to do the job. Usually the liquid waste from these high-tech (tertiary) plants treat their liquid waste or discharge with chlorine. The treated liquid is then dumped into a nearby lake or stream.

The argument is that dilution of the waste water is the solution to the pollution. Dumping a million gallons of waste water, for example, into the Hudson River—which contains billions of gallons of moving water and flows directly into the Atlantic ocean—can't be harmful. But it can and it is.

There was a time when there were so many sturgeon in the Hudson that they were called Hudson river beef. Now marine life in the mighty Hudson is just about gone. Lake Erie once was considered dead, and many other large and small, once-pure bodies of water are dead. It might be difficult to believe that man could possibly adversely affect the Pacific Ocean with his sewage, but we are doing just that. Sewage discharge is ruining much of the offshore kelp industry along the coast of California.

Imagine for a moment what could be accomplished if all human and animal waste were not wasted. If you add the minerals produced by humans to the minerals produced by the livestock and plants upon which we and our animals feed, you come up with the figure of 5 pounds of phosphorus and 26 pounds of nitrogen per individual per year. This is just about the quantity of fertilizer we spread over our fields every year.

For example, a city of 100,000 people and their livestock produce sufficient waste nutrients to grow 100,000 tons of plants a year. In other words, when wastes are not discarded, plants, animals and humans form a closed circle. They are energy sufficient unto themselves.

If we utilized our wastes, as they do in many other countries, we would not need to expend energy mining and otherwise producing fertilizer by the millions of tons.

The use of septic tanks or other home sewage treatment systems reduces the quantity of waste that would otherwise go into our public sewers, through the treatment plants, and into rivers and lakes. The solids that remain in a septic tank can be removed every few years and can be treated and used as fertilizer. It depends upon what your tank-cleaning company choses to do with it.

Our economy benefits because on a per capita basis a home sewage treatment system doesn't cost a fraction of what a municipal plant costs in equipment. And don't forget the cost of installing and maintaining the sewer lines. Home sewage systems require almost no energy (with the exception of a few specialized systems not described in this book). Central sewage treatment plants consume tremendous quantities of energy in one form or another—oil, gas, electricity.

BACK TO THE SEPTIC TANK

For many years, health officials believed that a central, municipal sewage treatment plant could dispose of human waste more effectively and safely than was possible with individual septic tank systems. Between 1950 and 1970, some 10 million homes discarded their individual tank systems and connected their soil lines to municipal sewers. Time proved them wrong. Not only were the costs phenomenal, but as our dead and dying rivers, lakes, and streams proved, the method did not eliminate pollution. The present trend is toward individual septic tank systems and centralized systems that operate on the principle of the septic tank. The main difference is merely size.

Man has always abhored his feces. It is smelly and vile and induces a natural revulsion. Human waste is an insidious, persistent source of infectious diseases. It is a kind of poison. Like all poisons, it doesn't take very much of it to cause harm; a trace can bring you down.

Being highly soluble, feces can be removed with soap and water. But its easy solubility multiplies its danger. Once in contact with water, that water becomes polluted and remains so for a considerable length of time.

In Ancient Times. Although the people of times long past gave considerable thought and care to the securement of what they believed to be pure drinking water, human waste was disposed of with no more care than their garbage.

In fifth century Athens, for example, human waste was simply permitted to pile up at the outskirts of the city. Hippocrates, the famous Greek physician (460 to 370 B.C.) wrote extensively about public hygiene and the importance of pure drinking water, but supposedly he wrote and taught nothing about the safe treatment of human waste.

The Romans built sewers primarily for the conveyance of storm water. Whatever else was carried along was of secondary importance. Most of the homes and apartment houses were served by cesspools or covered storage tanks behind the stairs. Every now and again, manure merchants would collect the waste and sell it as fertilizer. People who lived on the second and higher floors didn't always use the common, ground-floor privies. They would use chamber pots and dump the contents out of their windows and onto the streets.

There were practically no sewage systems in any of the cities and towns during the Middle Ages. Chamber pots were normally dumped in street ditches and open sewers. Once in a while, lime and carbolic acid was used to slush the gutters and open sewers. Outside of the towns and cities people used cesspools. Where there was running water, they positioned their privies so that the waste dropped directly into the stream. Medieval Paris was noted for its smell. The west side of towns and cities were favored because winds usually blow from west to east. The east side of town had the worst odors.

Epidemics Were Common. Under such conditions it is not surprising that entire villages and cities fell prey to hepatitus and typhoid. Many health officials believe that our recent phenomenal population growth is not due to advances in medicine but to advances in municipal hygiene.

Modern Sewage Systems. In 1543, the city of Bunzlau in Silesia was purportedly the first to install a sewage treatment plant. London did not build a sewer for human waste until 1815. From Roman times until about the 1840s, little improvement in sewers and sewage treatment was made. Mainly, waste was moved from one place to another, and little else was done to it.

The Septic Tank Arrives. Like many important inventions, the septic tank did not spring fully grown from the mind of one man. Two Frenchmen, Mouras and Moigno, found that a tank or box placed in a sewer line between a house and a cesspool, which the line fed, would trap the solids in the waste. The solids could then be removed from the tank. That was much easier than cleaning the pool.

This work was followed by a Scotchman named Donald Cameron. In 1896, he developed the modern sealed, (almost) airtight septic tank chamber in which anaerobic bacteria attack and destroy the pathogens in domestic waste water. All modern septic tanks follow this basic design.

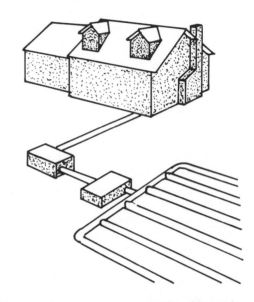

Chapter 15

How Septic Tanks Work

THE TERM SEPTIC TANK IS AN ABBREVIA-
tion for septic tank sanitation system. The
system consists of a tank and a drain field that work
together to cleanse and purify household waste
water. It is an imminently practical, safe, and effective system that is currently used by half the rural
homes in the United States and Canada and about 15
percent of urban American homes. In California,
legislation has been passed that encourages the
installation and use of individual sanitation systems.

Outside of a biannual cleaning and some care,
the system requires no energy and no attention.
The tank can last indefinitely. The drain field will
have to be treated or replaced after 20 years or
more. The system is invisible and odorless. It has
no moving parts. See Fig. 15-1.

The flow of sewage is by gravity from the
house to the tank where the sewage is stilled and
permitted to stand from three to five days. Here the
anaerobic bacteria break down the solids. The
liquid flows from the tank by gravity and is called
effluent. The effluent enters the drain or leaching

field where it seeps into the soil that filters the
liquid while aerobic bacteria break the liquid down
to nutrients and chemicals that support plant life.
When the effluent finally reaches an aquifer, it is fit
for human consumption.

BASIC DESIGN

The average individual with a moderate propensity
for cleanliness uses anywhere from 50 to 350 gallons per day for drinking, cooking, washing, and
flushing the toilet. With the exception of homes that
have provisions for separating the grey water from
the black, some 99.9 percent of the effluent is
water. The remaining 0.1 percent is solid. Of this,
roughly 80 percent is organic and the balance is
inorganic. The organic substances derive from
feces, detergents, soaps, urine, food bits produced
by garbage grinders, and food bits simply discarded
without care. Water softeners, borax, paint, photo
chemicals, household cleaners, and the like are the
sources of the inorganic solids found in household
sewer lines.

Roughly 50 percent of the water coming down

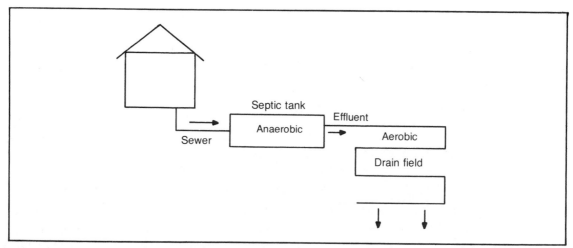

Fig. 15-1. The parts of a septic tank system.

the sewer pipe is only slightly polluted. This is water from the kitchen sink, the shower, and the bathtub, and it is termed *grey water.* The balance is water used to flush the toilet. Because it is heavily polluted, it is called *black water.*

In terms of quantity, there isn't much truly dangerous substances in the black water. On an average, humans excrete between 16 and 100 ounces of feces every day. This is dissolved and carried through the house sewer by 50 to 350 gallons of water per person per day. Thus we have a system that handles a large quantity of liquid with a small percentage of solids.

The Tank and Its Function. The tank is a large, watertight, airtight and lighttight container positioned beneath the earth. Sewage is brought to the tank by an extension of the house sewer line or pipe. A second pipe leads from the tank to the drain field. The field consists of a number of loosely joined pieces of clay pipe (drain tile) positioned within a bed of gravel and covered by a layer of earth. The earth provides support for the grass that is indistinguishable from any other grass on the property.

Water movement is by gravity alone. There is no odor whatsoever and the area can be used for any purpose excepting the passage of cars and trucks.

The tank acts as a stilling pond and as a breeding ground and incubator for anaerobic bacteria,

yeasts, fungi, and actinomycetes.

When the comparatively rapidly moving waste liquid leaves the sewer pipe and enters the tank, liquid movement almost ceases. This serves two functions. The solids sink to the bottom and the anaerobic creatures thrive as they grow much better in still water than in moving water.

The anaerobic community of microscopic creatures attack and digest the organic solids as they sink to the bottom of the tank. In the process, methane and other gases are produced as a by-product. The gas bubbles to the surface of the liquid, bringing along fine particles of solid matter. These fine particles, together with oils and grease, form a scum over the surface of the liquid. This scum further insulates the anaerobic creatures from whatever air seeps into the tank. The underside of the floating layer of scum provides an ideal breeding place for the anaerobic community.

So long as there is organic matter in the tank and its temperature does not drop below freezing, the bacteria continue eating, multiplying, and breaking down the organic material into its constituent elements. The anaerobic bacteria, yeasts, fungi and actinomycetes work comparatively slowly, but faster when they are warm and more slowly when they are cold. Roughly, they take 10 to 20 times as long to convert organic matter to soluble nutrients as do the aerobic creatures.

$$\text{Anaerobic creatures} + \left(\text{an organic substance}\right) \xrightarrow[\text{enzyme}]{} \begin{array}{c}\text{bacteria}\\\text{protoplasm}\\C_5H_7O_2N\end{array} + \begin{array}{l}CH_4 \text{ methane}\\H_2O \text{ water}\\NH_3 \text{ ammonia}\\H_2S \text{ hydr sulfide}\\PO_4 \text{ phosphates}\end{array} + \begin{array}{c}\text{40-60}\\\text{Kcal}\\\text{(energy)}\end{array}$$

Fig. 15-2. Sequence of organic changes.

In doing their thing, the anaerobics produce only about one-tenth to one-fifteenth the heat as do the aerobics. Note that the anaerobics do not completely transform organic matter such as feces into its component elements. That job can only be preformed by the aerobic creatures in the presence of sufficient oxygen.

Attacking an organic substance, anaerobic creatures produce the sequence of changes shown in Fig. 15-2. Physically, the change is from a solid to a liquid. What the anaerobic community cannot digest—pieces of stone, plastic, comparatively large pieces of bone and wood—remains in the tank and must eventually be removed by pumping. With a good-sized tank and a little care, pumping need not be required more than every other year.

The *effluent,* the liquid that flows by gravity alone out of the tank, is still septic. That means it contains substances that promote the decomposition of vegetable and animal matter as well as pathogens, bacteria, and viruses harmful to man and beast. The effluent cannot be simply dumped. It must be treated and it is treated so simply and naturally that it is a wonder that it took man so long to discover and apply the method. Very simply, let the discharge water from the septic tank percolate through 4 feet of soil and it becomes pure enough to drink. Nobody knowingly depends on a mere 4 feet (authorities differ on this distance) of soil to purify drinking water. Generally, a minimum distance of 60 feet is required. The effluent should not be discharged into the earth less than this distance from a well or spring. There are other considerations in the placement of a drain field in relation to a water supply that must also be considered. They are discussed farther along.

THE WORLD UNDER OUR FEET

We stand on it, we lie down on it, and sometimes we even roll around on it. Soil is soil to the naked eye, and with the exception of the earthworm, groundhogs, and the like, many people regard the soil as uninhabited. This perception is entirely false. A fertile acre can house half a ton of worms. If you bothered to count, you would find some 5 million bacteria in a single teaspoon of ordinary soil (Fig. 15-3). Viruses, the smallest creatures found in the soil, are about 0.02 microns in size and they enter the bodies of the microbes and eat their protoplasm. This, of course, is not good for the microbes, but is not necessarily bad for our purpose.

When times are very good for bacteria, when the temperature is high and there is plenty of air and food, they multiply like crazy and form thick gelatinous colonies in the soil. The gelatin reduces the flow of liquid through the soil, thus reducing the effectiveness of the drain field. By killing off some of the bacteria, the viruses slow the growth of the bacterial colonies. The bacteria are also destroyed by the amoebas, which can be described as microscopic jellyfish. The amoebas are preyed upon by the nematodes that will eat both amoebas and microbes indiscriminately. The amoeba and other single-cell protozoa can consume more than 1 million bacteria a day. Generally, they confine themselves to about 100,000 microbes a day.

The nematodes are the largest of these creatures, and they are just visible to the naked eye. Put a dish of vinegar out in warm weather and in a few days you will see that it is swarming with nematodes.

Balance. The earthworms make comparatively the large holes in the earth. The nematodes make the smaller holes. Together, the worms and the nematodes aerate the soil.

The bacteria feed upon and digest the organic waste present in the effluent. In effect, the aerobic bacteria "cook" and chemically oxidize their food. Their wastes are soluble, stable chemical compounds that are harmless to man and beast and food to plants. Without the bacteria, fungi, and the rest of the microscopic creatures of the soil, plants would soon starve to death. Only those nutrients that could be leached out of the soil would be available.

This is what has happened on almost all American farms. The fields are completely cleaned of all plant remains after the crop has been harvested. There is no longer much humus (organic material) in the soil. There is little microbial activity in the soil and the farmer has to feed his plants. He does this by applying artificial fertilizer.

Aerobic creatures attacking the effluent from a septic tank produce the sequence of change shown in Fig. 15-4. Although the heat produced by the digestive activity of the aerobic bacteria is comparatively slight, it is sufficient to encourage the

Fig. 15-3. Top: A stalked, free-swimming ciliate; a kind of protozoon. The large black specs are bacteria, drawn roughly to scale. The small dots are viruses. They are not to scale. Microbes range to about 5 microns. Viruses invade and eat microbes. Bottom Left: A nematode is just large enough to be barely visible to the naked eye. Bottom Right: An amoeba is a kind of microscopic jelly fish that lives on bacteria, but falls prey to the larger, tougher nematodes.

growth of plants. Nature has set up a symbiotic balance between man and the animals, the microscopic creatures in the soil, and the plants. The creatures in the soil work to purify waste and convert it and dead plant matter into food for the plants.

Human Health. The community of creatures naturally living in our soil destroy pathogens (disease-causing bacteria, viruses, and the like) in six ways.

☐ Soil bacteria, fungi, and other microbial organisms produce antibiotics such as penicillin that destroy pathogens.

☐ Temperature, acidity, and moisture within the soil is so different from that found in human and animal bodies that the pathogens soon die.

☐ Protozoa in the soil prey upon bacteria and the viruses along with the bacteria.

☐ Soil organisms compete with pathogens for food, thus starving them out.

☐ The soil acts as a filter and prevents the larger bacteria from traveling very far.

☐ The soil—especially the clay—absorbs the viruses and locks them in place, preventing them from moving on.

FILTRATION

The effluent from the septic tank is led by means of a perforated pipe or small sections of pipe with openings between sections into the drain field. The liquid flows out of the pipe and through the coarse gravel in which the pipe rests. From the gravel, the liquid passes into the soil. Some of the liquid will leave the soil by evaporation (Fig. 15-5) and some of the liquid will leave the soil by transpiration—a process wherein the roots of the plants take up water and the water evaporates from the plant's leaves. In this way, plants act to draw water from the soil. The balance of the effluent moves through the soil, generally in a downward path.

The soil acts as a filter, but simply passing septic water through several feet of soil will not physically filter out anything but the larger particles. A purely physical filter will not remove viruses, but soil does because its action is far more complex and effective than a mere physical filter.

Soil is mainly composed of tiny pieces of stone and humus (decayed organic matter such as leaves, etc.). When the pieces of stone are very fine, the soil is called *clay*. When the particles are a little coarser, it is called *silt*. Coarser still and it is *sand* followed by *gravel*.

The visual difference between types of soil is considerable. Soil types can be distinguished fairly easily by visual examination and feel. The actual difference is far greater. Clay particles are typically less than 1/12,500 of an inch across. Fine silt ranges from 1/12,500 to 1/500 of an inch across. Fine sand ranges from 1/500 to 1/250 of an inch. Medium sand particles may measure 1/100 to 1/50 of an inch across. The smaller the particles of soil the greater their total surface area. One pound of soil has approximately 1 million square feet of surface area. That is an area as big as the entire state of Connecticut.

The Electric Charge. Every particle of stone in the earth carries a minute electric charge that performs great and wonderous service. All pathogens have a coat of protein. In most soil, this coating somehow produces an electric charge of a polarity opposite to that of the particles of soil. Because pathogens are submicroscopic viruses, they cannot resist the pull of the electric charge on

$$\text{Aerobic bacteria} + O_2 \left(\text{effluent}\right) \xrightarrow{\text{enzyme}} \begin{array}{l} \text{bacteria} \\ \text{protoplasm} \\ C_5H_7O_2N \end{array} + \begin{array}{l} CH_4 \text{ methane} \\ H_2O \text{ water} \\ NH_3 \text{ ammonia} \\ H_2S \text{ hydr sulfide} \\ PO_4 \text{ phosphates} \end{array} \quad \begin{array}{l} 40\text{-}60 \\ \text{Kcal} \\ \text{(energy)} \end{array}$$

Fig. 15-4. Sequence of organic changes.

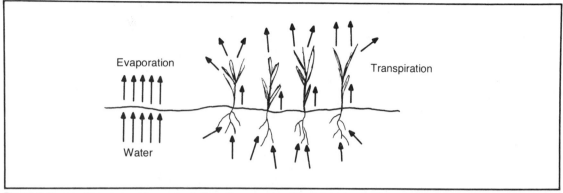

Fig. 15-5. Evaporation occurs when water in the soil evaporates directly into the air. Transpiration occurs when water in the soil evaporates from the leaves of a plant into the air. Because the roots of a plant has a tremendous total length and surface area, and a plant has considerable leaf exposure to the sun and air, transpiration can account for easily a thousand times more water evaporation from a given soil area than the surface of that soil's area alone.

the particles of stone. They are attracted to the stone and held in place. Pathogens away from their hosts die fairly quickly.

The finer the particles of stone the greater their total surface. Clay, which has the greatest surface area, develops so much attractive power that pure clay will cleanse water of whatever viruses might be present in less than 4 inches of travel. In other words, pass polluted water through 4 inches of clay and it will be purified and fit to drink.

In addition, the electric charge of clay and all larger particles of stone to a lesser degree precipitate such dangerous chemicals as strontium, rendering it harmless, and react with proteins and nutrients so they will not wash away and react with lime to change soil acidity.

Two points need clarification. No one depends on just 4 inches of clay to purify the effluent from a septic tank. Drain fields are at a minimum positioned 50 feet from a water supply. If there was nothing but clay between the drain pipe and the well, you would have 150 more times the filtering action than you required at a minimum.

Second, clay alone cannot be used as a filtering medium. It packs too tightly; it is not porous. The rate of water movement through pure clay is so slow you can form a bowl out of wet clay and use it to hold water. Put another way, you cannot install a drain field in a field of pure clay; water won't leave the drain pipe.

The septic tank household waste disposal system consists of a tank and a drain field. Sewage from the residence flows by gravity into the tank where it remains for three to five days. During this period, heavy particles settle to the bottom and anaerobic bacteria attack and digest the organic material in the sewage. The sewage, now converted to a liquid, flows by gravity into the drain field, which consists of a series of perforated or separated sections of pipe positioned beneath the earth in a bed of gravel.

The effluent flows into the earth where aerobic bacteria convert the remaining organic material to soluble, stable plant nutrients. In flowing into and through the earth, the effluent is filtered. Large particles are stopped by the grains of clay and sand. Viruses are absorbed onto the surfaces of the particles of stone and remain there to die. The now purified water can find its way downward to an aquifer or it can return to the surface by means of evaporation or transpiration.

During the life of the system, and this can be 20 to 30 years and more, there is no odor. From the surface, the drain field looks just like any other section of lawn. The tank is also out of sight. Except for the need to pump it every year or two, its life span is unlimited if it is made of concrete or brick.

The life span of the drain field depends upon its use, the size of the field, and the nature of the soil in which it is positioned. When the soil immediately surrounding the drain pipe becomes saturated and clogged with fine particles of solid matter, the soil loses its ability to function as a filter. The effluent cannot flow easily out of the drain pipe and disperse.

As a result, the effluent backs up and appears on the surface of the ground, bringing with it an odor. Therefore the septic tank system has a built-in alarm that warns the home owner of trouble. Note that the appearance of water and even an odor above the drain field is not always a certain indication that the field is saturated; other conditions can also bring this about.

When a drain field has reached the end of its useful life, it can be restored and made an effective filter again by replacing the gravel and treating the adjoining earth.

Chapter 16
Selecting a
Septic Tank System

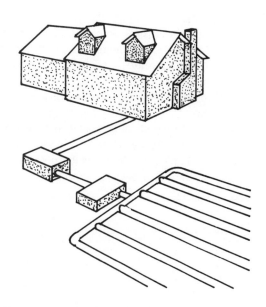

THE PURPOSE OF THE TANK IS TO STILL THE FLOW of sewage from the building and contain the sewage long enough for the anaerobic bacteria to do their work. Estimates are that it takes the anaerobic bacteria from three to five days to digest the fecal matter and other organic substances in the sewage. Therefore the tank must, at a minimum, be large enough to properly contain three days worth of sewage. An oversize tank does no harm (except to one's pocketbook). An undersize tank fills up more quickly and requires pumping more often.

If the drain field has been sized to meet the requirements of the smaller tank, and the tank is overloaded, the drain field will be overloaded. Some raw sewage will flow into the drain field. Black scum might appear on top of the drain field. The scum consists of anaerobic bacteria and undigested sewage. It will have a very unpleasant odor. Pools of water might also appear on top of the drain field. Water appears on the surface of the earth because the undersized drain field does not permit proper evaporation and transpiration.

Because the tank will contain a tremendous quantity of pathogens, the tank must be perfectly watertight. It must be impossible for any of the contained liquid to escape to the surrounding soil.

Because the action of anaerobic bacteria produce an odor and a number of gases (mainly methane, which is flammable), the tank must be airtight. To prevent the tank from bursting in response to the pressure developed by the methane and other gases produced by the anaerobic bacteria, the tank must be vented one way or another.

Because no one wants to go to the trouble of replacing the tank every few years, the tank must be made of some type of rot- and rust-resisting material.

Because the most practical place for the tank is underground, the tank must be strong enough to support a layer of earth above itself, and the traffic that passes over it.

SELECTING THE MATERIAL

Tanks are commonly made from any number of

materials. Your problem is to find the best compromise between price and life span. Costs will depend upon whether or not you construct the tank yourself, if you purchase a ready-made tank, the type of material selected, and your distance from the supplier. Also, the weight of the tank requires consideration. A precast concrete tank, for example, will require special equipment for transportation and positioning. Two men can handle a fiberglass tank.

Redwood. Redwood has an average life span of 30 years. Can be constructed on the site by the homeowner. It needs to be coated with asphalt on the inside. It rots most quickly at the waterline (from the outside in). It cannot be positioned too deeply in the earth because the dirt load on top

becomes a problem and bracing becomes expensive.

Fiberglass. The life span of tanks made from fiberglass is given as 30 years. Fiberglass consists of glass fibers impregnated with plastic. Plastic is nonbiodgradable. No one is really certain whether fiberglass tanks placed in the earth will last 30 years or longer.

Fiberglass tanks are manufactured in a number of sizes, and a unit large enough for the average home weighs no more than 300 pounds. Two men can easily position the tank within the excavation. Make certain the tank is strong enough to support the load of soil. See Fig. 16-1.

Metal. The most practical metal from the viewpoint of cost would be mild steel. The most

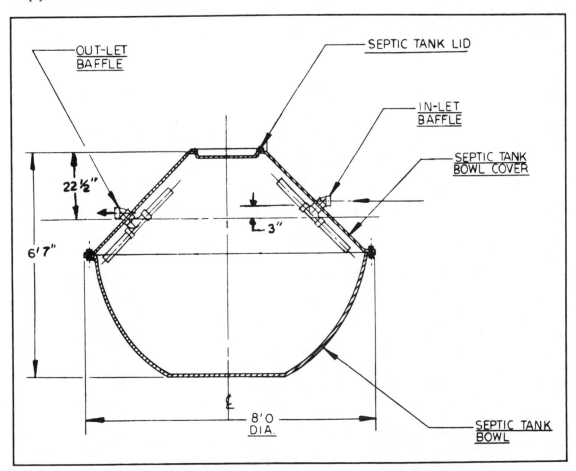

Fig. 16-1. A fiberglass septic tank. This one holds 1,000 gallons, and is easily lifted by two men.

practical method of fabrication would be by electric arc welding. The tank could be fabricated in a shop and hauled to the site on a truck, but you would need a heavy-duty crane to lower the tank into position. The expected life span of a steel tank would be 7 years or less. Its insides should be given a coating bitumastic tar to extend its life span.

Cast Concrete. The cast concrete tank is made by constructing a wood form and filling the form with concrete. The concrete might or might not be reinforced with steel bars. It all depends on the size of the tank and its cover. The life span given for concrete tanks is 20 years, but I believe this figure to be far less than the actual life span of a well-constructed concrete septic tank. The inside of the tank should be coated with tar.

Concrete Block. There are two types of concrete block: lightweight and heavy. The light block is sometimes called *cinder block*. It should not be used for the walls of a septic tank. When the blocks have been positioned (laid up), the interior walls of the tank should be given two ⅜-inch coats of cement plaster. When the plaster has hardened, it should be covered with a layer of bitumastic tar.

The life span of a concrete block septic tank is usually given as 20 years. If its inner surface is cement plastered and then coated with tar, the tank should last years longer.

Brick. The life span of septic tanks constructed of brick is probably a little less than that of concrete block. But like the block tank, the brick tank's useful life can be extended by plastering and tarring.

DRAIN-FIELD PIPE

Unglazed Clay. Sometimes called *terra cotta*, these are short sections of pipe usually 4 inches in diameter (internal) with either a circular or an octagonal exterior. They are positioned in line with about ⅛ to ¼ of an inch of separation. The space permits the contained liquid to flow out of the pipe and into the surrounding earth. A strip of tar paper placed over the top of the joint keeps the soil out until the soil compacts and bridges the opening.

The useful life of the tile cylinders is given as 20 years, but this figure appears to be very low.

Terra cotta has survived from Roman times and even earlier.

Plastic Pipe. Plastic pipe is a lot easier to install than the clay pipe. Lengths to 20 feet are usually available. Joints are made with the aid of plastic cement. A little plastic cement is daubed on the end of the pipe and the pipe slipped into its fitting. About 30 seconds later the joint is permanent. The plastic pipe used for drain fields is perforated with two rows of holes on opposing sides of the pipe. In position, the pipe is rotated until the holes are horizontal (that is, they are at the sides of the pipe).

The life span of plastic septic tanks is given as 30 years. There is no reason to believe the plastic drain field pipe will not last as long.

SIZING THE TANK

It is important that the tank be large enough to accommodate the flow of sewage that will enter it during an average three- to five-day period. The difference between three and five is 66 percent. This is a considerable difference in size, costs, and labor. Unfortunately there are no firm guidelines available for favoring either three-day capacity or five-day capacity. The only suggestion I can make is that the thermal location of the tank should influence your decision. Bacteria work much more slowly as the temperature drops. The farther north you are the larger the tank should be.

Oversize. Also bear in mind that an oversize tank has its advantages. The effluent will remain for a longer time in the tank so there will be a greater degree of digestion and breakdown in the sewage. Consequently, the liquid entering the drain field will have a smaller percentage of solids.

Because it is the solids in the effluent entering the drain field pipes that are a major factor in eventually clogging the drain field and forcing the homeowner to renew the field, the "cleaner" the effluent entering the pipes the longer the field will last. All septic tanks have to be cleaned from time to time. The solids the anaerobic bacteria cannot digest sink to the bottom of the septic tank and collect there. Eventually, there is so much solid matter in the tank, that tank capacity is considerably reduced and the solids could even be high enough to block the flow of the sewage in and the effluent out of the tank.

The tank should be cleaned well before this occurs. In most home septic tank systems, the need for cleaning follows a regular pattern. Perhaps it is needed once a year or once every 18 months. When the tank is large, the time span between cleanings is increased. The septic tank pumping company charges per visit and not by the quantity of muck they remove. The oversize tank can save you money by extending the useful life of the drain field and reducing the number of clean-outs required.

Undersized Tanks. The initial cost of an undersized tank is obviously less than that of a similar but larger tank. Undersize, unfortunately, has no definite measurements. You can only be certain the tank is undersize when it results in problems. Some of the problems might not appear for years. If, for example, your drain field clogs up in 10 years instead of 20 or more, you can ascribe the premature failure to an undersized tank.

If you find you have to have your tank pumped every six months rather than every year or year and a half, you can be certain the tank is undersized. If the earth above the drain field is frequently wet, if the field emits an odor, if black muck appears on the earth above the field, you can be fairly certain the septic tank is undersized. This assumes that the field has been properly constructed and is of a proper size.

Codes. Many municipalities have building codes that specify the size of the septic tank on either the basis of bedrooms or head count. No matter what method the authorities use for sizing your septic tank, you have to abide and make your tank at least equal in size. There is no restraint preventing you from making your tank larger than the specified minimum. Some building departments will also specify the position of the tank, its method of manufacture, input and output piping, and possibly other details. Don't try to circumvent the building department. It might not appear so, but they are working for your benefit.

Bedroom Versus Head Count. The old method of determining required septic tank capac-ity was by counting bedrooms. This is fallacious. A bedroom could sleep one, two, three, four or even five. A two-bedroom house might in time be converted to a four-bedroom house. A more accurate way of estimating the quantity of water that will flow out of a house sewer is to count the number of people living in that house. When you count people, you must also count everyone over the age of five. Children bathe, shower, eat, and have their clothes washed. And if you want to be prepared for the onslaught of time, you need to assume that the infants now present will eventually grow up. In addition, more people might join the family.

Realistic Estimate. Chapter 2 explains how to estimate the total quantity of water the average individual uses per day. The figure comes to 95 gallons per person per day. This includes all water used for cooking, washing as well as water used in flushing the toilet.

Assuming there are five people in the household, water use and water flowing out the house sewer would amount to:

$5 \times 95 = 475$ gallons

Three days would bring the total to:

$3 \times 475 = 1425$ gallons

Five days would bring the total to:

$5 \times 475 = 2375$ gallons

When you multiply it all out, the two figures are very obviously considerably different. Which to use? To be certain, it is best to go with the larger-volume tank. On the other hand, if you are in a warm climate and have no reason to believe your family will grow larger, you might be safe going with the lower figure or compromising on a volume somewhere in between.

REDUCING THE SIZE OF THE TANK

There is a practical way of reducing the required size of your septic tank and associated drain field without reducing its efficiency or safety. Two types of water flow out your home's sewer pipe. They are mixed, of course, but by definition they are separate. The water that comes from the toilet containing feces and pathogens is called black water. The water that flows from the tub, dishwasher, shower, and lavatory is called grey water. Black water must be treated by anaerobic bacteria, aerobic bacteria, and filtration through the earth before it is safe enough to join the groundwater of the earth. Grey water can be disposed of by one or two methods. Note that while kitchen water is also grey water, it requires slightly different treatment.

Division of Waste Water. According to the National Water Well Association, water consumption in the average American home breaks down this way:

Total
Bathroom: 75%
Kitchen: 25%
Bathroom
Toilet flushing: 45%
Bathing: 30%
Kitchen
Dishes and laundry: 20%
Cooking and drinking: 5%

Reducing Septic Tank Volume. Some 45 percent of all the water used in the average household goes to flushing the toilets. This is black water. It must go through the septic tank and drain field. The other 44 percent is a combination of dishwasher water, bathwater and shower water. This water is classified as grey water. Grey water does not have to be treated because it ordinarily contains no pathogens. Dishwashing water contains oils and greases that should not be mixed with bathwater.

By separating the black water from the total volume of water flowing out of the house sewer pipe and into the septic tank and limiting the flow to the black water alone, the required safe size of the tank and field can be cut in half. This is accomplished without reducing the efficiency of the tank or the field. Other steps, of course, have to be taken for handling the bath and kitchen water.

Further Water Savings. Some of the popular, so-called low-flush toilets use only 3.5 gallons of water per flush. There is a savings of some 30 to 40 percent in water, which is also a savings if you are on a water meter (buying your water). The Japanese manufacture toilets with two flush systems. One for feces and the other for urine. The quantity of water that passes with each operation depends upon the direction you swing the control handle. Some models have a lavatory atop the toilet tank. When you wash your hands you produce grey water. This water goes into the toilet tank and is used for flushing.

Some homeowners have connected their tubs, washing machines, lavatories, showers, etc., to their toilets so that grey water is used for flushing. To do this, it is necessary to install some sort of automatic pressure valve to make certain there is always water of one kind or another in the toilet tanks, and to relieve the grey water flow should the toilet tanks be full.

In addition, you must make certain the traps in the various fixture lines are readily accessible. Grey water always contains a quantity of hair. In time it clogs the trap and must be removed.

Disposing of Grey Water. There are two ways of disposing of grey water other than running it through the toilet and eventually through the tank and field. One method is legal everywhere. The other method, called the Mexican drain, is prohibited in most parts of the country. The first, acceptable method consists of directing the water underground. The second consists of merely dumping it on the surface of the earth.

Grey water is used on commercial airlines. They color the water blue and use it to flush their toilets. Sud-Saver clothes washers saves the water used for the first washing and recycles it for use for the second washing. The Grand Canyon National Park, and perhaps other parks use grey water for toilet flushing. Campbell Soup saves the water it uses for washing its soup vegetables and directs it to its fields to irrigate crops.

In many parts of the country, grey water can be used to water trees, bushes, and grain crops. Used this way, the grey water does not come in contact with the edible portions of the plant. In other words, it is permitted to water your lawn with grey water, but you cannot use the same grey water to irrigate celery or lettuce. The grey water would come in contact with the edible portions of the plant and thus result in pathogens possibly being ingested. It doesn't matter that you run a greater risk of infection by kissing your girlfriend or being present near someone who sneezes, the government is not taking any chances with grey water.

UNDERGROUND DISPOSAL

Very simply, the grey water is led off to a *dry well*. In doing so, you have to bear two considerations in mind. The dry well must be large enough to absorb the grey water, and the pipe leading from the building to the dry well must either be well insulated or sufficiently deep in the earth to avoid freezing.

Sizing the Dry Well. Assuming that by now you have computed the sewage volume that will flow from your home—and you have decided whether to use the three-day, five-day figure or a volume somewhere in between—you can take 50 percent of this to find the grey water volume (and reduce the black water volume by an equal sum). You now have the required volume in gallons per period needed for the dry well. You know that, if you construct a well with this capacity, this is the quantity of liquid it will hold. You do not at this time know how *fast* the water will leave the well and enter the soil.

For example, if you are in coarse sand or gravel, liquid will leave the tank and enter the earth as fast or faster than it enters the dry well. In such conditions, you can reduce the size of the dry well safely. At the other extreme, you might encounter solid clay. Its rate of absorption is very low and you will have to construct a large dry well and probably alter the nature of the adjoining soil by adding a generous quantity of sand.

Chapter 20 describes the techniques that can be used to measure the porosity of the soil and estimate the rate at which it will absorb water. Refer to that chapter.

Begin with the total estimated figure of the quantity of water that will be used by the householders and what, if any, possible increase might likely occur with time. Divide this figure in half to secure the estimated total quantity of grey whater that is to flow into the well. Kitchen water will not flow directly into the dry well, but it will eventually reach the dry well.

Assume that we have a home with five people living in it. Each individual uses 95 gallons of water per day. This makes for a total water flow into the house of 475 gallons a day: $475 \times .50 = 237$ gallons of grey water per day.

Let's up this figure to 250 gallons/day to provide a small margin of safety. Thus you need a well with a total liquid capacity of 250 gallons (at a minimum). To convert this figure into feet of cubic volume, multiply: $250 \times 0.1337 = 33.4$ cubic feet. A tank with this volume will hold 250 gallons of water.

To find the dimensions of a tank capable of holding 250 gallons, the easy way is to find the cube root of the cubic volume that will encompass 250 gallons. This figure is 33.4 cubic feet.

$$\sqrt[3]{33.4} = 3.5 \text{ approximately}$$

Thus, a cube-shaped tank 3.4 feet on a side will hold close to 250 gallons. This appears to be the size of the tank you need, but hold on. You are not out of the well as yet.

When projecting your need as satisfied by a 33.4-cubic-foot tank, it is assumed that the tank would be completely empty until filled with grey water. It is also assumed that the water would be completely absorbed by the surrounding earth at the rate of one tank full a day. Neither of these two assumptions hold true in all cases.

Types of Dry Wells. There are currently three types of dry well designs. Eventually, some company might introduce a plastic dry well. The available well types can be categorized as tank, positioned block, and rock pile.

The tank design provides 100 percent volumetric efficiency. No interior space is lost. The positioned block design is about 75 percent efficient and the rock pile's efficiency is about 33 percent. If you construct a tank, you can build to an inner tank dimension of $3.5 \times 3.5 \times 3.5$ feet to secure the 250-gallon volume needed. If you go to positioned block, you have to increase the tank's inner volume by at least 125 percent. If you select the rock pile, you have to increase the overall volume by at least three times.

The type of dry well constructed affects the overall volume of the dry well. This is one factor and it assumes that the surrounding earth is capable of absorbing the grey water at the rate of one tank full a day. Obviously, this is not always the case. Absorption rate should be measured, as described in Chapter 20, and the resultant figure should be used to determine the necessary exterior area of the tank—less its top. Only the area of the sides and the bottom are considered areas of absorption.

Building a Tank Well. Tank wells (Fig. 16-2) are true tanks in that they are completely empty inside and have masonry walls with holes on their sides (none on their bottoms).

Excavate the required hole, making the sides as vertical as practical. Level the bottom and cover it with a 4- to 6-inch layer of gravel. Level the gravel as best you can with a rake. Build the walls in the conventional manner using 10-inch cinder blocks spaced ¼ to ½ of an inch apart. Use no mortar between blocks but do use mortar between courses (rows).

Build the walls straight up until you approach the last 2 or 3 feet, and then begin to corbel the block inward. You do this to reduce the size of the opening at the top of the dry well. Using cement mortar, plaster the exterior, horizontal joints of the corbeled block. This is done to strengthen the corbeled portion of the dry-well wall. As you work your way up, use a stock to keep mortar from slipping into the joints between adjacent blocks. Fill the open space behind the block with coarse gravel.

The top of the well is terminated at least 1 foot below grade and the opening closes with a slab of 3-inch bluestone or flagstone. For strength, all the joints in the topmost course should be mortared, and the cinder block's tops and exterior surface should be mortar plastered. This will keep the block on top, where the frost can penetrate, from absorbing water. See Fig. 16-3.

Positioned Block. Use 10-inch cinder block and any cinder block pieces you might have. The blocks are placed, hole side up, in more or less a circle around the perimeter of the hole. The space between is loosely filled with more block. The second course of block is positioned atop the first. The game plan here is to fill the entire hole with as

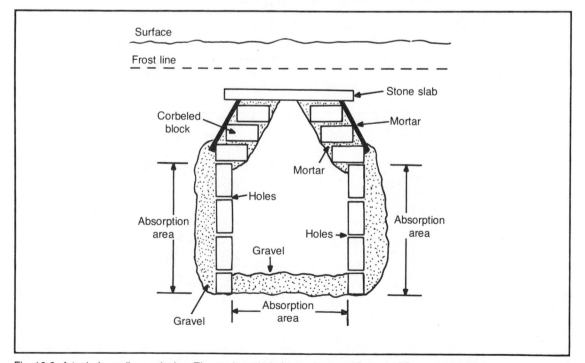

Fig. 16-2. A tank dry well, one design. The top is corbeled to make a small stone slab cover possible.

175

Fig. 16-3. The spaced-block dry well. The blocks are spaced so as to form a wall and to support a flat rock or concrete slab cover.

few blocks as possible, still retain the dirt wall, and still provide a stable support for the slab of bluestone or flagstone that will support the earth roof.

Roughly, you will use at least twice as many blocks with this design as with the tank design, but

you do not have to trouble yourself with mortar and actually laying brick. On a small well, this method is practical. As the well size increases, the ratio of block to well size increases at a tremendous rate and the design becomes too expensive.

For best results, the space between the block and the dirt wall of the excavation should be filled with coarse gravel. Top the block with flat stones and tar paper. See Fig. 16-4.

Constructing the Rock Pile. You excavate a hole in the earth and simply fill it with quarry-run rock. These are the leftover stones produced by blasting at the quarry and then removing all the stones of a desirable shape and size. Quarry run is the least expensive stone commercially available, but it is just what you want: ill assorted, odd shaped, and varied in size. They are just the kind of stones that will not pack evenly but will form large voids when dumped into the hole.

If you scrounge for stones, go for the large stones with awkward shapes. When you dump them into the hole, try to keep them from packing and lying evenly.

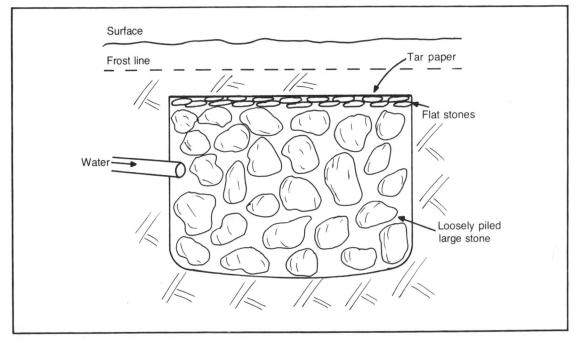

Fig. 16-4. The rock pile dry well. Fieldstone or quarry rubble piled helter skelter in the hole, with the flat stones saved for the top to provide support for the tar paper that holds out the soil until it compacts.

The 33 percent efficiency figure is derived because odd-shaped stones in a pile usually have voids between them amounting to one-third the total volume of the pile of rock. When you toss these rocks into a hole, two-thirds of the rock will be in direct contact with the soil. There will be no transference of water from the rock to the soil. This will only take place between the voids and the soil.

To complete the rock pile well, place flat rocks and gravel on top of the topmost layer of stone to provide a more or less flat surface. Cover the surface with two layers of tar paper. Then cover the paper with soil. Pile sufficient soil on top to form a low mound. In time, the soil will compact and it will all be level. If you just add enough soil to make it all level, in a short time you will have a hollow there.

The Best Type. The tank well is far and away the best choice. It has a greater liquid capacity for its size than the others, and it can be cleaned out with a reasonable amount of labor. The other two types require all the stones and block be removed from the hole before it can be cleaned. That is a tremendous, troublesome job.

GREY WATER PRECAUTIONS

Grey water contains phosphorus and nitrogen in varying quantities. Few plants can take grey water as a steady diet and thrive. Therefore, grey water should not be used exclusively on any plants. Alternate with fresh water.

If you are running the grey water through a hose and onto your lawn or your garden, cover the end of the hose loosely with a cotton bag so that the water is filtered. Change or wash the bag often.

Don't let grey water contact crops such as cabbage, spinach, lettuce, etc. On other plants, you can use grey water if you alternate with fresh water to prevent pollutant buildup. Use grey water for hosing down the drive, washing the car, and the like.

TREATING KITCHEN WATER

Most homeowners do not attempt to separate the kitchen sink and the dishwasher drain water from the black water. This is a double mistake. The waters flowing out of the sink and out of the dishwasher contain oils and grease. Neither oil nor grease is easily digested by the anaerobic bacteria in the tank. Most of the oil and grease will float to the top of the liquid in the tank to form a scum. In time, this scum will combine with laundry detergents and enter the drain field. The muck soon clogs the soil. Running the kitchen and dishwasher drain water in with the black water does not induce overnight drain-field failure, but it does definitely shorten to its useful life. See Fig. 16-5.

Grease Trap. The proper treatment for the kitchen and dishwasher drain water consists of running it through a grease trap. Traps can be purchased ready-made from a plumbing supply house or fabricated from metal or concrete. Essentially a trap consists of a container having a 1- or 2-cubic-foot volume and one or more baffles—plus a watertight, removable cover. The greasy grey water flows in one end, is slowed by the baffles, and then flows out again.

During the period the greasy water is within the trap, the water is cooled. The grease and oil settles out or floats to the surface. The grey water leaving the trap is essentially free of oils and grease. You can use metal for a grease trap because the grease coats the metal and prevents corrosion. See Fig. 16-6.

The trap should be positioned in a cool but not cold spot. You want the liquid to be chilled, but not frozen. Every year or so the trap's cover should be lifted so the grease can be removed with an old spoon or stick. To save on piping, you can run all of the black water through a grease trap. The method works, but you expose yourself to an awful smell each time you have to clean the trap. Some feces will always collect there too.

SEPTIC TANK DIMENSIONS

You have counted heads and you have decided on the capacity of the tank in gallons. You now have to convert the gallonage to cubic feet. A cubic foot contains 7.48 gallons. Dividing the required gallon capacity of your tank by 7.48 gives you the required cubic volume of the needed tank.

Because the cube is not the most practical shape for a septic tank, you cannot approximate its

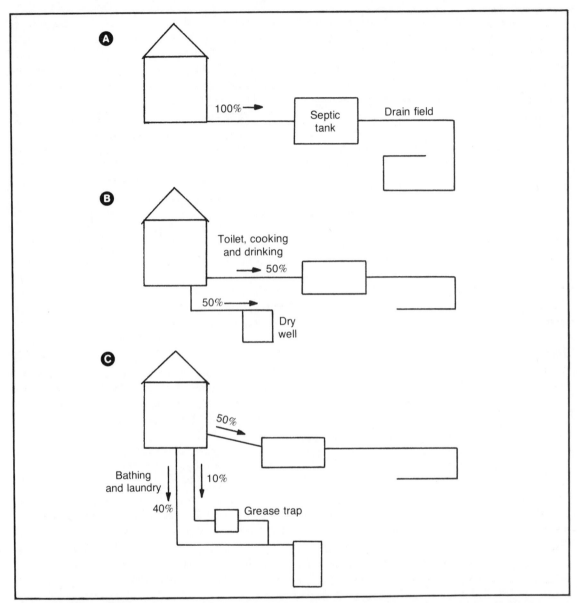

Fig. 16-5. A: The total flow of sewage out of the house enters the septic tank. B: Here the sewage is divided into black water (50%), and grey water (50%). The grey water is led to a dry well, thus cutting the load on the septic tank and drain field in half. C: To keep the dry well from being clogged with cooking oils and grease, the kitchen water is first directed through a grease trap before entering the dry well.

size by simply finding the cube root of the required cubic volume and adding a couple of feet in height to provide the necessary clearance above the level of the black water. The easy way is to use Table 16-1. This table provides for tanks about 25 percent over

capacity. It is not at all necessary to make any tank to the exact dimensions given.

DESIGNING THE SEPTIC TANK

Assuming that you have decided to construct your

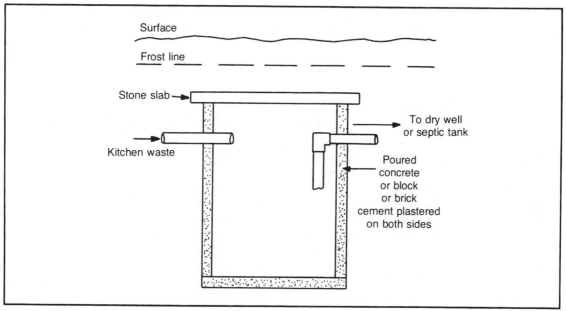

Fig. 16-6. Basic arrangement of parts of a grease trap.

own tank rather than purchase a ready-made tank, your next step is to decide on the type of tank you want to build. The five to choose from are the:

☐ Single compartment.
☐ Double compartment.
☐ Triple compartment.
☐ Meander.
☐ Siphon.

As you go down this list, the tanks become more complicated, somewhat more expensive to construct, a little more difficult to maintain, but more effective in removing the solids from the effluent. The siphon type also provides a somewhat more forceful flow of the effluent from the tank into the drain field.

Maintenance Comparisons. The single tank has a single trap door and access opening. In the usual course of events, the pump crew opens

Table 16-1. Minimum Recommended Septic Tank Dimensions.

Number of people	Appox. gallons	Inside width	Inside length	Liquid depth	Total depth
4	250	3' 6"	3' 0"	4' 0"	5' 0"
6	360	4' 0"	4' 9"	4' 0"	5' 0"
8	480	4' 0"	5' 0"	4' 0"	5' 0"
10	600	4' 0"	6' 0"	4' 0"	5' 0"
12	720	4' 0"	7' 3"	4' 0"	5' 0"
14	840	4' 0"	7' 9"	4' 6"	5' 6"
16	960	4' 0"	8' 9"	4' 6"	5' 6"

Recommendations based on a figure of 95 gallons per day per person, with 40% of the effluent being directed to a dry well (57 gallons of black water into the tank per person per day.) Note: If a garbage disposal unit is to be installed, tank capacity should be increased 50%, assuming the kitchen sink will drain into the septic tank.

the trap and loosens the sludge at the bottom of the tank with a long-handled shovel. A large-diameter hose is then admitted to the tank and an external pump removes the sludge. In a double-compartment tank, this operation has to be repeated because there are, in effect, two tanks.

With the three-compartment tank, the cleaning operation has to be repeated three times. In the meander tank, the number of times the access doors or traps have to be opened depends upon construction. With most designs, it will be necessary to open at least three or even four ports. The siphon design usually comprises three compartments so there is a need for three ports.

Single-Compartment Design. Figure 16-7 shows the basic components of the single-compartment septic tank design. What is important is that you use a 4-inch or larger input pipe and an output or drain pipe of the same internal diameter.

Use a T on the input side and seal the top end of the T with a piece of bronze screening. The top of the T should be an inch or two below the top of the tank. The T permits gas produced by the anaerobic bacteria in the tank to escape via the sewer pipe and either out the fresh-air vent or up the soil stack.

The outlet pipe can be either an elbow or another T. It should be positioned on a plane with the inlet pipe. If the outlet is lower than the inlet, the maximum practical depth of the tank will not be utilized because the liquid will drain as soon as its level is above the outlet pipe. If the outlet is higher than the inlet, the liquid in the tank will rise above the inlet pipe, creating an undesirable back pressure.

Use either vitrified clay pipe for the inlet and outlet or plastic pipe with sanitary fittings. Sanitary fittings have no inner ribs. This reduces the possibility of muck catching and remaining there.

Two-Compartment Design. The total volume remains the same. Inlet and outlet pipes are positioned just the same. The only change is a divider plus two ports. The divider is positioned so as to make the outlet compartment about 40 percent as large as the inlet compartment. The divider need not be watertight.

In a masonry tank, it can be fashioned of block

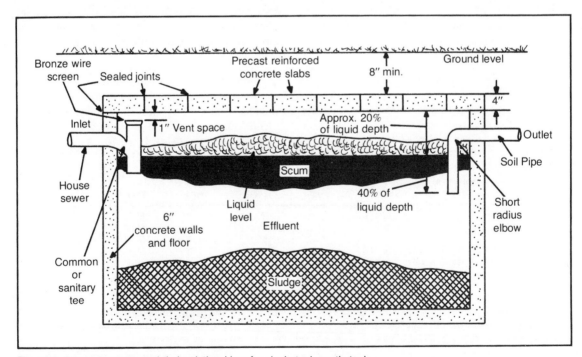

Fig. 16-7. The major parts, and their relationship, of a single-tank septic tank.

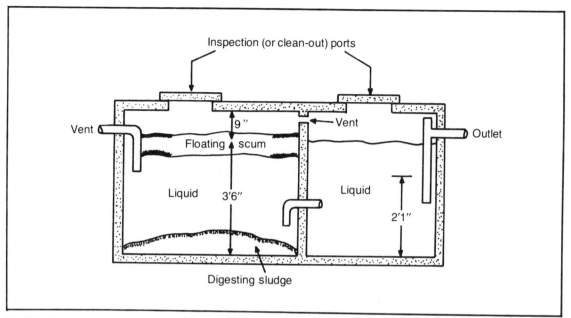

Fig. 16-8. A two-compartment septic tank. Dimensions are given just to illustrate the approximate sizes of the parts. The vent is there just in case the elbow between tanks becomes clogged.

laid on edge. In a wooden tank it can be a partition made of wood. You can use either a short length of pipe to connect the two compartments or you can leave an opening approximately 5 inches square in the dividing wall about three-fourths of the way up the side of the wall.

The purpose of the divider is to slow the movement of liquid even more than it normally would when entering the single tank. Slowing the movement of the liquid does two things. It gives the microbes more time to do their thing and it helps settle the solids. This keeps more solids out of the drain field where the solids will, in time, clog the field.

Because there are two compartments and no easy way of reaching one compartment from the other, you need two trap doors or ports. While the partition will keep most of the solids out of the second compartment, in time you or the tank crew will have to get a shovel down into it to loosen the sludge.

Another feature you might find advisable to add is a *relief vent* in the divider. This is a 5-inch-wide, 2-inch-high hole at the very top of the divider.

Its purpose is two-fold. It permits the movement of gases from one section to another. It permits the liquid to flow from compartment to compartment should the communicating pipe somehow become clogged up. See Fig. 16-8.

Three-Compartment Tank. Again, the requirements for the placement of the inlet and outlet pipes and their construction remains the same. The compartments can be divided on the volume basis of 2:1:1. The ratio need not be exact, but it is important that you increase the overall inside dimensions of the tank to make up for the space given to the partitions.

Again, you need a relief valve or aperture at the top of each compartment divider, and you need ports and access to each of the three compartments.

The three-compartment tank doesn't do anything the one-, and two-compartment tanks do. It just does it better. It keeps the effluent clearer and gives the bacteria even more time to work on the sewage.

The Siphon Tank. The siphon tank is a bird of a different color. The design can be used with one-, two-, or three- compartment tanks. The original had three compartments. See Fig. 16-9. With the siphon septic-tank design, the sewage flows into the first tank and then the second tank by siphoning from the first. Flow into the optional third tank is also by siphon.

The operation is shown in Fig. 16-10. It works this way. When the level of the liquid in the second compartment rises above the elbow of the siphon, water begins to flow up this pipe and down into compartment three. The pipe or the siphon is now filled with liquid and the liquid continues to flow until the level of the liquid in section two falls below the entrance or end of the siphon. Then flow ceases and no flow resumes until the liquid level in section two is once again higher than the siphon's elbow. Then the cycle resumes.

The advantage of this arrangement is pressure. In the previously described septic tank designs, the rate of liquid flowing out of the tank is never greater than the rate flowing in. When, for example, a toilet is flushed there is a momentary gush of possibly 8 gallons of water. The water, now sewage, enters the tank and possibly raises the level of the liquid already in the tank by a fraction of an inch. It is this fraction of an inch that flows out of the tank.

Phrased hydraulically, the head, the pressure on the effluent, is less than 0.1 psi (pound per square inch). The flow out of the tank is therefore a mere trickle. It is not backed by any pressure; it has no speed.

With the siphon arrangement, the third tank is normally empty. When the liquid in the first tank rises high enough to start the siphon action, the liquid pours from the first tank into the second tank at a prodigious rate. A 5- or 6-inch pipe should be used for the siphon. The third tank is quickly filled and emptied. There are a few moments when the third tank is nearly full.

Assuming that the liquid is 4 feet deep at this moment, and that the liquid, the effluent, is drawn off at the lowest possible point from compartment three, there will be a 4×0.43 head on the effluent (or a pressure of 1.72 psi). This pressure is many times more than possible with any other non-powered arrangement. At this pressure, although it is comparatively low as water pressures go, it will still drive the effluent with considerable force and speed. This force will help the liquid reach all parts of the drain field and carry along any obstructions present in the drain-field feeder pipe.

Unfortunately, the siphon design has one basic limitation. In computing the *head* that will be developed (head meaning pressure on the liquid), it is assumed that the drain pipe will be positioned at the very bottom of compartment three. When the drain has to be positioned near the same level as the intake pipe, there is no point to installing a siphon. The higher you have to position the drain or outlet pipe, the less head that is developed. Head is directly dependent upon the height of the water above the drain and is computed at the rate of .043 pounds per foot of height.

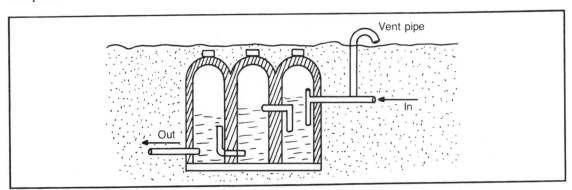

Fig. 16-9. A three-compartment tank designed in 1883 by Edward S. Philbrick of Boston, Mass. The air vent helps relieve gas pressure in the tanks, but hinders the anaerobic bacteria. The siphon provides a fast-flow action from the in tank at the right to the middle tank.

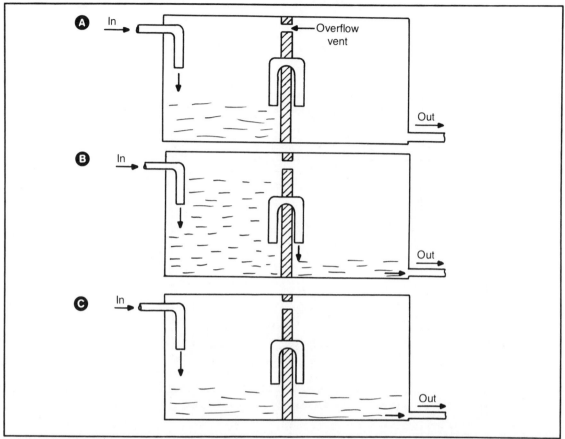

Fig. 16-10. How a siphon tank works. A: Black water enters at left. The siphon is empty. B: The in tank fills up to a point above the top of the siphon. Water now flows down the right-hand siphon leg. C: Water continues to flow down the right leg until the water in the in tank is about level with the water in the out tank. In this way, it is possible to have a succession of half-tank fulls of effluent in the out tank that will provide some pressure; far more than the slow entrance of the sewage into the in tank.

The Meander Design. Supposedly, the septic-tank design shown in Fig. 16-11, perfected by Dr. T.J. Winneberger, retains more solids and produces a clearer effluent than the other designs previously described and illustrated in this book.

The design was devised to force the effluent to meander like a slow-moving river. The slower the stream moves, the more sand and silt it drops. The more turns the river makes, the more it drops going around the turns. By installing the partitions lengthwise in the tank and forcing the liquid to travel in comparatively long paths and make several turns, the time the effluent spends in the tank is greatly increased.

The first partition has a hole about halfway up

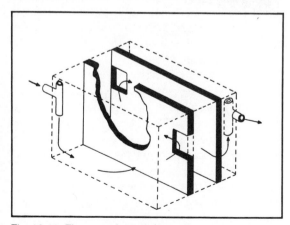

Fig. 16-11. The meander tank forces the sewage to take a long, slow path through the septic tank.

183

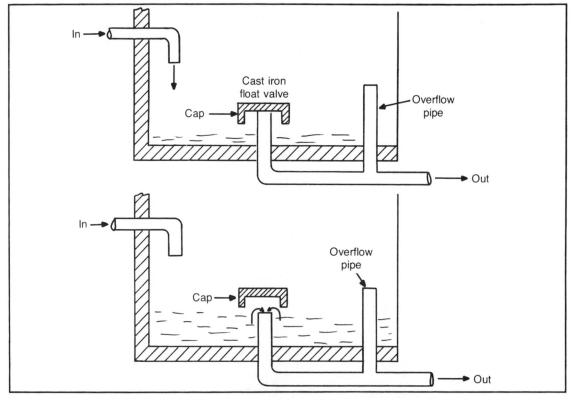

Fig. 16-12. A simple, automatic valve produces some surging action in the outflow of the effluent in this design.

its side. The second partition has a hole at the opposite end near its top edge. The drain is positioned at the end of the tank, away from the second hole or opening. Thus the liquid has to travel not only back and forth, but also to some degree up and down. In a tank that is 6 feet long, the path from the inflow pipe to the outflow pipe or drain pipe is about 22 feet.

The meander tank has three compartments so it must have three ports for access and cleaning. The partitions can be of any material that will not rust nor rot easily. The partitions need not be watertight. In a masonry tank, concrete block can be used for the partitions and the holes are provided by simply omitting several blocks where needed.

Automatic-Valve Design. Still another septic-tank design worth mentioning is the automatic-valve design. Very simply it consists of the usual watertight roofed container with the usual

inflow pipe. Two pipes take care of the outflow. One is for normal operation. It comes up through the bottom of the tank and rises vertically to a height a short distance below the lower end of the inflow pipe. This pipe is open at its upper end.

The vertical, main outflow pipe carries a heavy, cast-iron cap. The cap is inverted, open end down. When the tank is empty or the liquid level is below the cap, the cap's weight seals the outflow pipe. When the liquid level rises above the cap, the air under the cap raises the cap and the liquid can enter the outflow pipe.

Thus there is a series of outflow surges that are interspersed by periods when the black water is entering the septic tank. The second outflow pipe is also vertical, but it extends well above the at-rest position of the top of the cap. Should the cap become stuck for any reason, the liquid can flow out the second, overflow pipe. See Fig. 16-12.

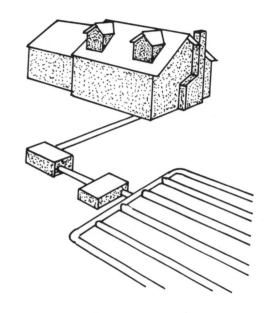

Chapter 17

Layout

THIS CHAPTER DESCRIBES POSITIONING THE septic tank and the drain field. In a few cases, the positions occupied by the tank and field naturally are ideal for the purpose. In many more instances, however, positioning the tank and the associated field is always a compromise. It is important to know what are the factors that must be considered and what weight should be given to each factor.

DRAW A MAP

The best way to plan the layout of your septic tank and drain field is to work with a topographical map. It is not difficult to draw one.

Drawing the Plan View. If you have a copy of the *plot plan*—the blueprint that "came" with the property, the print drawn by a licensed surveyor—you have all the major dimensions. You can transcribe this plan onto graph paper. That will make it easy to keep the dimensions in their proper relationship. If you do not have a plot plan, go to work with a steel tape and measure the grounds. In

either case, add on the building and whatever else is present on your property (barn, shed, etc.). See Fig. 17-1.

Draw a map of your land and your planned drain-field location. Beware the zoning laws that protect your neighbors. Keep clear of drives, walks, large shade trees, and the shady side of the house. Position the field downhill from your and your neighbor's well and at least 60 or more feet away. Some authorities caution having an even greater distance.

The figures at the right side of Fig. 17-1 are contour lines. As you can see, the road is considered to be at zero elevation, and the area immediately adjoining is fairly level and at the same elevation (0).

As you go away from the road, you see a line reading 1.0 feet. That means that area, as indicated by the contour line, is 1 foot higher than the road. Farther in the same direction, the next contour line is marked 2.0 feet. The ground along this line is 2 feet higher than the road, and so on. Reading the

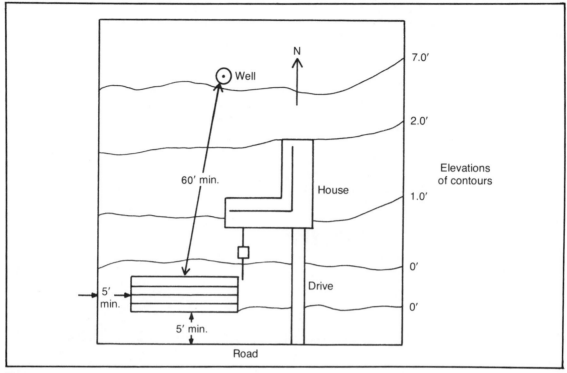

Fig. 17-1. A map of the drain field.

contour line figures, you can see that the property ascends more quickly toward the rear, reaching a height of about 7 feet.

Adding the Topography. Drawing the topography is not difficult because it does not have to be exact. Assuming that there is a pitch to the land, start with the highest point that is convenient. Drive a short stake into the ground. Attach a line to this stake. Hold one end of the line in your hand. Hang a line level from the line. Stretch the line. Lift the line until a helper tells you the level's bubble is centered—meaning the line is level. Measure down from your string hand to the earth. This distance is the elevation at this point.

Note the elevation on your map. The peg represents 0, hand-to-earth distance elevation. Repeat at a number of points. On the map, draw lines connecting points of equal elevation. See Fig. 17-2.

Now reverse the elevations. The peg is the highest point. The other points are lower.

With your map in hand, you can move your

septic tank and drain-field positions about until you find the best position for them. With a topo map, it is far easier to visualize the various factors that must be accounted for than it is to do the same walking about on the property.

When and where the building department or the local health authorities have a say in your tank and drain field, you might be required to bring visual evidence of where you plan to place everything before the authorities will approve.

BEST DRAIN-FIELD LOCATION

Given a choice, the best location for the drain field is to the south side of the building, away from the shadow of the building and possibly shade trees and bushes. In this position, the drain field will receive the most sunshine.

The drain field may be placed beneath a lawn, but it should not be placed beneath a vegetable garden or a tree or bushes. The drain field should not be located in an area subjected to flooding or in

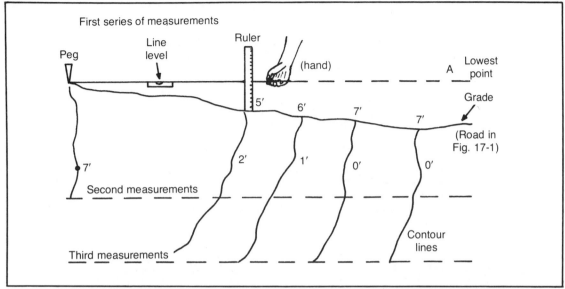

Fig. 17-2. How contour lines are secured. With a few simple tools, elevations over an area are ascertained and then transferred to a scale map of the same area.

an area that is excessively wet. In the latter case, steps must be taken to "drain" the drain field.

If there is an option, the drain field should not be placed on the side of a hill because construction there is difficult. The drain field should be positioned downhill from the nearest well because groundwater tends to follow the slope of the earth's surface.

DRAIN-FIELD DIMENSIONS

The dimension of the total length of drain pipe in the field is computed on the basis of liquid flow per time period and the rate at which the soil can absorb the liquid and dispose of it by transpiration and evaporation. This is discussed in Chapter 21. The design of the field, whether the drain pipe will form one serpentine snake or divide into a number of arms, or whatever will affect the overall space the field will occupy. Here are the general rules that apply.

☐ The outer wall of one drain field trench to the next should be equal to 1½ times the effective depth of the trench. The effective depth is the distance from the center of the pipe to the bottom of the trench.

☐ On slopes, the effluent must be prevented from "running away," meaning running quickly out of the higher pipes to flood the lower pipes. In this way, the lower pipes become quickly clogged while the upper pipes remain relatively unused.

☐ The field should not be positioned where an automobile or a truck might accidentally back onto it. The weight of the vehicle will probably crush the pipes.

☐ Septic tank and drain field must be set back (separated) a minimum distance to satisfy good neighbor policy and the local building department.

Septic Tank Elevation. The elevation of a septic tank is crucial. At best it can only be a few inches off. The tank should be deep enough to permit 1 foot of cover. See Fig. 17-3. In cold climates, this depth can be increased to 2 or more feet. The inlet pipe is centered in a hole about 8 to 10 inches below the top edge of the tank.

This is to provide sufficient room for a pipe T and allow a clearance of an inch or two for venting. The feed pipe hole can be moved up or down a few inches, but no more. Lowering the pipe reduces the capacity of the tank. Raising the pipe reduces or eliminates the vent. Here comes the crucial part.

187

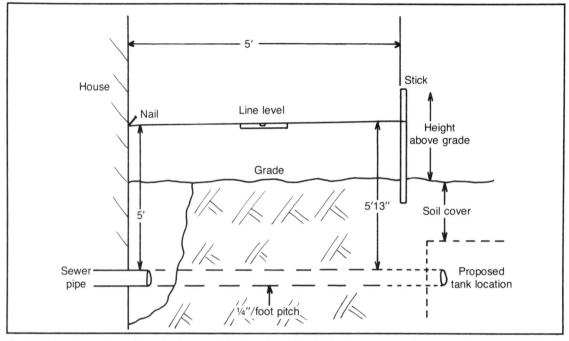

Fig. 17-3. Positioning the tank at the preferred elevation in relation to the house sewer pipe.

The sewer pipe running from the house to the tank can be run at an angle or pitch of ¼ to ½ inch to the foot. Less of a pitch and the muck will hang up. Increase the pitch and the liquid will run off, leaving the solids. The pitch has to be held within these figures (¼ to ½ inch to the foot), or it can be run at an angle of 45 to 90 degrees.

Bear in mind that the sewer pipe is not flexible. You can "bend" it at the joints a fraction of a degree, but no more. If you try to bend the joint any more, the joint will leak. You can, of course, secure angle fittings, but they are limited to 22½, 45, and 90 degrees. The elevation of the septic tank must therefore be such that you can make the sewer pipe run at either a slope of ¼ to ½ inch to the foot or incorporate one of the commercially available elbows.

Checking Elevation. You can measure up from the top surface of the sewer pipe as it exits the building on the inside of the building and transfer this dimension to the outside of the building. A more accurate method consists of digging a short trench at the outside of the building down the side of

the foundation until you reach the sewer pipe. Then measure up from the top edge of the sewer pipe any convenient distance and mark this distance off with a nail driven into the side of the building.

Let's say you have measured up a distance of exactly 5 feet. Next you go to where you plan to excavate for the septic tank. Drive a stick vertically into the earth at the spot where you believe the side of your tank should go. Stretch a line from the nail in the side of the house to the stick. With the aid of a line level, raise or lower the stick end of the line until the line level indicates that the line (string) is perfectly level.

If, as in my example, your nail is 5 feet exactly above the top surface of the sewer pipe, you now know that you have to go down exactly 5 feet from the end of the line to make the sewer-to-septic tank pipe perfectly horizontal. But you need a pitch. With a steel tape, measure from the side of the house to the stick. This gives you the necessary length of the pipe run (not exactly, but close enough).

Next, multiply the pipe run in feet by ¼ inch.

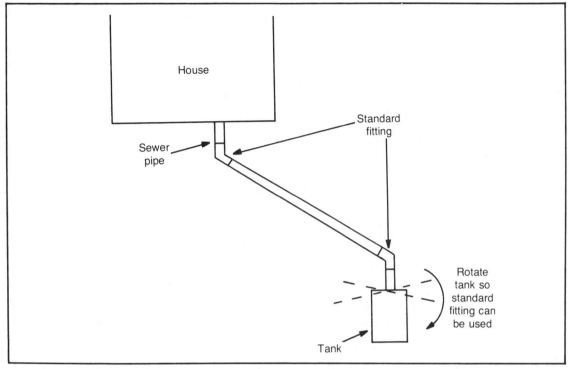

Fig. 17-4. The importance of rotating the tank so that standard angle fittings can be used in the sewer pipe line.

This will give you the drop necessary at the tank end to produce the desired pitch. Let us say the run is 52 feet.

Pipe run in feet × ¼ inch = required drop
52 feet × .25 = 13 inches

Bear in mind that the nail is 5 feet above the sewer pipe. The tank end of the sewer pipe has to be 5 feet, plus 13 inches below the horizontal line.

Now you can measure down from the stick end of the line to determine the elevation of the surface of the soil (the grade) at that point in relation to the house sewer pipe. That will tell you how much soil there will be above the tank.

Knowing the vertical dimension of the septic tank you plan to construct or install and knowing where you want the sewer pipe to enter, you can calculate the necessary elevation of the bottom of the tank based on the distance that will separate the tank from the house (the length of the sewer pipe).

When you actually excavate for the tank, you can double check your figures with the aid of a vertical stick or pole positioned where the sewer pipe will enter the septic tank. Bear in mind that you are working with the top surface of the pipe. To find the center of the hole you will need, you must take half of the overall diameter of the sewer pipe and add it to the measurement.

TANK ROTATION

The inlet T fitting that you will install will come through the tank wall at a right angle. When at all possible, that T should point directly at the house sewer pipe as it leaves the building. If at all possible, you want no turns in the sewer pipe from the house to the tank. Turns slow the sewage down and can provide hang-up points for the sewage. See Fig. 17-4.

If you have to go around a corner or make a turn, you have no option but to do so. Bear in mind that you have but a limited number of angle fittings with which to do so. They are 22½, 45, and 90 degrees. If you cannot point the side of your septic

189

tank directly at the house sewer pipe, you must position it at an angle that will "work" with one of these standard fittings.

One way to do this is to draw the building and the tank to scale on graph paper. Using a protractor, determine what fitting can be used and the angle of the tank side. Another method is to lay out the pipe "dry" on the surface of the ground. Use the fitting you believe will work and replace the side of the tank with a straight board.

Distance. Whether or not the distance from the end of the house sewer pipe to the facing side of the septic tank is crucial or not depends upon the material you are going to use for the connecting pipe. If the sewer connection is to be made with PVC (plastic) pipe, there is no problem. You can easily cut the pipe to length with a saw. Extensions of any length can be easily made with a slip connec-

tor. That is not so with cast iron or vitrified clay. Both are joined by means of a bell and spigot, usually, and both are difficult to cut.

The major problem that could arise is that you will end up short by a few inches. You cannot cut the bell end of cast-iron pipe down to a few inches in length, and neither can you do the same with the bell of clay pipe.

If this condition arises, you will have to open the joints and shorten a few of the preceding lengths of pipe so that the piece can be at least 6 inches long. By measuring with a steel tape carefully from the house end of the sewer pipe to where the septic tank wall will be positioned, you can determine if that is the correct distance or vary it a few inches as necessary. Or you can make the pipe run "dry" and see where the last section of pipe falls. See Fig. 17-4.

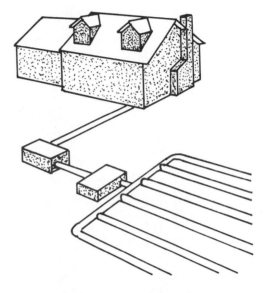

Chapter 18

Tank Construction

YOU CAN PURCHASE A READY-MADE SEPTIC tank and have it deposited within the excavation you have provided, or you can hire a contractor and have a tank constructed in whole or in part on your property. To find a tank manufacturer you can check your local telephone directory or look in *Thomas*. The latter is a multivolume set of books that lists all major manufacturers by product and by location. Most libraries have these reference books on hand. As for finding a contractor, any general building contractor should be capable of constructing a tank, but you will probably also need to hire a plumber to make the pipe connection (union or building code regulations).

BUILDING YOUR OWN TANK

Depending upon your locality, it might or might not be required that the building department approve of and inspect your septic tank. They might or might not insist that a licensed plumber make the sewer pipe connections. In any case, it is best to play safe and check with your building department before starting construction.

Types of Construction

Chapter 16 describes various septic tank designs. This chapter covers construction materials and how they are used. Whether you choose to make a single-unit tank, a double-unit tank or even a triple, construction material and methods remain the same. Except for details, what follows applies to all designs.

Choice of Materials. For the average do-it-yourselfer, the choice of materials practical for the construction of a septic tank is limited. You can use redwood, brick, or concrete block and poured concrete. It is difficult to say which is the least expensive material for a given tank size and design, and difficult to say which requires the least amount of labor or skill. It all depends upon your experience, the cost of material when you build, and where you live. You cannot estimate labor very easily, but you can "price" the materials with just a little effort.

In the case of redwood, you have to construct a six-sided box backed by 2 × 4s every 12 or 18 inches. The 2 × 4s must project 4 inches beyond each edge so that each piece of 2 × 4 is 8 inches

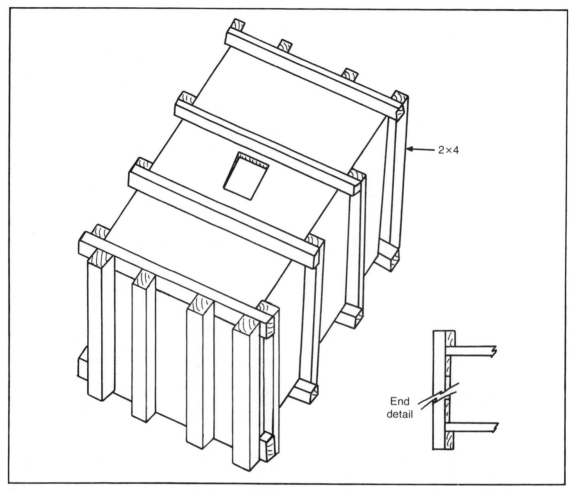

2×4

End
detail

Fig. 18-1. One way an all-wood tank can be constructed. The framing on the outside leaves a smooth interior.

longer than the width and the height of the box. See Fig. 18-1. In addition, you need 1-×-6-inch, tongue-and-groove sheathing sufficient to cover the entire tank's six sides. Bear in mind that when you assemble 1 × 6, T&G board you lose about 1 inch per board. Thus, three 1 × 6s side by side will not cover 18 inches; they will only cover about 15 inches.

To compute the amount of concrete needed to pour a tank, you simply find the volume of the concrete and then multiply by 110 percent to make up for the concrete lost when it sticks to the chute, the shovel, or falls to the ground. If you use a single-sided form, you have to roughly double the

quantity of concrete you need. Unless you have old boards lying about or have a use for concrete stained lumber (it can be used for subflooring), the wood needed for the forms is an additional expense.

To compute the cost of a concrete block tank, first figure the cost of pouring the slab foundation. Then find the total exterior wall area of the tank in square feet. Multiply this number by 144 (the number of square inches in a square foot). Now divide by 128 (the number of square inches in an 8-×-16-inch block). This will give you the number of blocks you need for the walls.

In addition, you will need a number of 4-inch solids, blocks to build the separating walls (if you

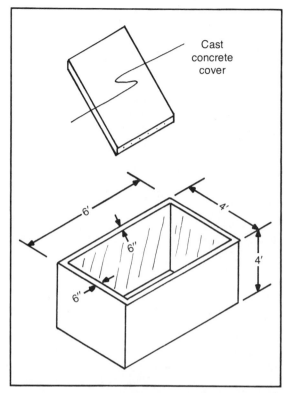

Cast concrete cover

Fig. 18-2. This concrete block or poured tank with the dimension shown will have an approximate total capacity of 67.5 cubic feet. This makes it capable of holding a total 504.9 gallons. It gives the tank an approximate working capacity of some 400 gallons.

are going to use them). You will also need sufficient concrete to pour the cover, steel bars for reinforcing the concrete, or precast concrete slabs with which to make the cover. See Fig. 18-2.

Septic tanks with walls more than 4 feet high or so should not be constructed with walls that are but a single brick thick. They will be too weak. Instead, use two bricks laid flat, side by side. Using standard (common brick) laid up in mortar joints ½ an inch thick, you will need 124 bricks to every 10 square foot of wall area, and about 2 cubic feet of mortar to do the job.

Skill

Not too much skill nor experience is required to work with any of the materials mentioned. Neatness doesn't count because the tank is out of sight.

Squareness and exact dimensions are not important either. Required tank capacity is at best only an estimate. Just the tank's depth in the earth and its orientation in regard to the house sewer line required particular care. Other than that, just make it strong and watertight.

Excavating

The decision to be made in regards to excavation is whether to go at it with a pick and shovel or hire a backhoe to do the job. Even a small hoe can do the job in half a day. If you are out of condition, digging the hole could take several days. Some of the excavated dirt can be left to one side. The balance can be hauled off. If the cost of hauling doesn't appeal to you, you can simply spread the soil over your grounds. Wait until you see how much has to go back in the hole, beside the tank, and how much the dirt over the tank will subside with time.

The depth of the excavation will depend on the height of the tank you plan to build, plus gravel, plus 1 foot of cover soil and even 2 feet in cold areas. The angle of the walls will depend upon the depth of the hole and the nature of the soil. In clay, you can dig almost straight down and the clay will remain in place (if it doesn't get water soaked). In soft soil, you will have to angle the walls to prevent them from collapsing in on you and your work.

To construct a wooden tank, you have to make the bottom of the hole 10 or 12 inches larger in both directions to provide the clearance you will need for the studs. For a concrete block tank or a brick tank, you need not make the hole much larger than the exterior dimensions of the planned tank.

If you are going to pour the tank and the soil is soft, you have to construct a double-wall form. The outer form is backed by studs and walers. See Fig. 18-3. Thus the outermost portions of the form will project 10 inches or so beyond the wall to be poured, in all four directions. This means the hole for the poured concrete tank with a double wall has to be at least 24 inches longer and wider than the proposed tank.

If the soil is very firm and the tank height is reasonably low, it is sometimes possible to use a single-wall form for a poured concrete tank. See

Fig. 18-4. The concrete is poured between the form and the earth wall. Perhaps twice as much concrete is used this way, but the material and labor of the second form is avoided.

BUILDING A WOODEN TANK

Start by making the bottom of the excavation flat and level. Cover the bottom with 4 to 6 inches of medium-size gravel. With a rake, spread the gravel and make it as level as possible.

Making the Bottom. Cut the 2 × 4s that will form the bottom frame (Fig. 18-5) to length; that should be 8 inches more than the desired width of the tank. Place them in position on the gravel. Lay

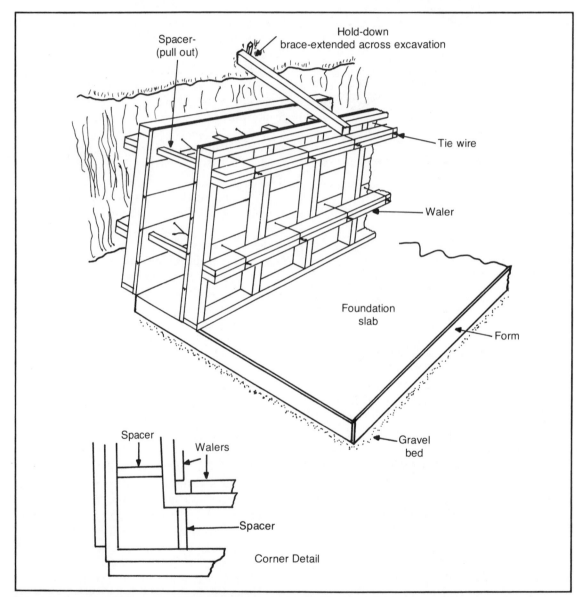

Fig. 18-3. How one wall of a poured concrete tank can be framed. Note that the bottom, the supporting slab, has already been cast. Corner framing is shown in the top view.

194

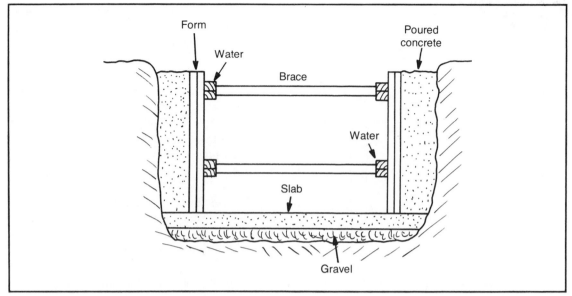

Fig. 18-4. When the soil is firm, a single-wall form can be used and the concrete slowly poured between the form and the soil. Note the supporting slab was cast before the wall forms were erected.

them on edge and space them about 16 inches apart as measured center to center. The distance from the outside edge of one 2 × 4 to the farthest 2 × 4 should be exactly equal to the desired inside length of the tank.

Remove the tongue from one length of 1-×-6 board. Lay this across the studs and just 4 inches in from their ends. Nail this board to the studs with 8-penny, common stainless steel nails, the kind used for making house foundations of wood. Position and nail the balance of the floor boards in place. Before you nail the last one, make certain its edge terminates 4 inches in from the ends of the studs.

Making the Sides. If you are a good carpenter, you can construct the sides outside the excavation, lower them into place, and nail them fast—one after another. If you are not experienced, the following methods will probably produce better results.

Cut the vertical studs to the necessary length. Then cut the upper horizontal studs (the studs that will carry the roof or top). These studs should be just as long as the studs lying on the ground.

Nail a pair of vertical studs to one of the studs that will carry the roof to make a U. See Fig. 18-6.

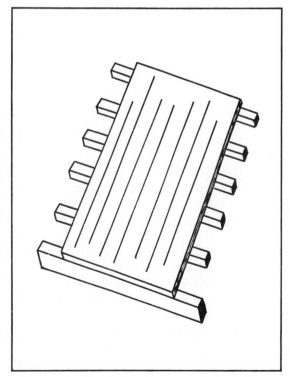

Fig. 18-5. One way the bottom of a wood tank can be constructed.

195

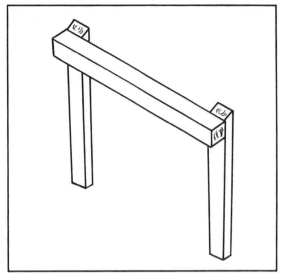

Fig. 18-6. One rib that will hold the tank's sides and top.

Repeat this operation until you have one U for every stud lying on the gravel.

Turn the Us upside down. Nail the ends of the Us to the ends of the studs on the gravel. See Fig. 18-7. To hold the Us in their proper vertical relationship, you will have to fasten a brace across all the top sides of the inverted Us.

Now you can line the long sides of the tank with 1 × 6s. If you find that the studs tend to spring back when you nail them, have a helper back them up with a second hammer. Just have it held against the 2 × 4.

Making the Ends. Fasten a vertical 2 × 4 at each end. Use a short one so that the corner timbers are doubled. Position the others 8 inches apart on centers. Fasten each end 2 × 4 with just two stainless steel nails. Use nails 6 or more inches long, one nail to each end. Drill a hole for each nail. Drive the nail through the hole, and then with pliers and hammer, clinch the nail. Bend it back upon itself. Now you can fasten the 1 × 6s to the vertical end studs.

Making the Roof or Top. The boards are nailed to the underside of the roof studs. No studs are on the inside of the tank. All the studs are on the outside. Leave an opening for cleaning and inspection. Block the opening with 2 × 4s. See Fig. 18-8.

You can make the cover of redwood fastened to a 2 × 4 frame or you can use a slab of 3-inch bluestone or flagstone. Its weight will hold it in place.

Sealing the Tank. With a stick or small spatula, fill all the corners with a generous quantity of asphalt. Then lock the asphalt in by forcing a length of quarter round into each corner and nailing the strip of wood in place.

Next, give the entire inside of the tank a generous coating of asphalt cement (not tar). This will help seal and preserve it.

Making the Connections. The weak point in a wooden tank is where the pipes enter and leave. The tank wall is only ¾ of an inch thick. Any movement of the pipes tends to open these joints. One way to prevent this is to increase the thickness of the tank at the point of entry and exit. To do this,

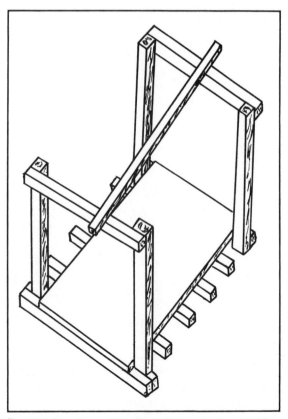

Fig. 18-7. Two ribs positioned and a temporary brace holding them in place.

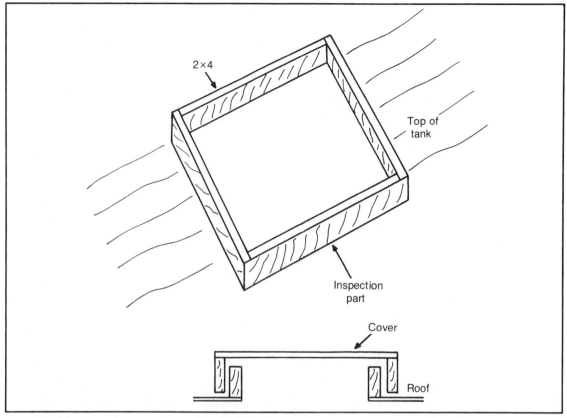

Fig. 18.8. How a port can be fashioned in the top of a wooden tank.

nail a 2-×-10 section on the inside and another on the outside where the hole in the tank will be. Use a saber saw to cut the circular hole (after drilling a pilot hole). The wall thickness at this point becomes close to 4 inches. See Fig. 18-9. Seal the joint with mastic, the non hardening plastic used for weatherstripping and window caulking.

BUILDING A CONCRETE BLOCK TANK

The hole is excavated as suggested and its bottom is covered with 4 to 6 inches of medium gravel or crushed stone. The gravel should extend a few inches on each side beyond the planned exterior edges of the tank. Make the surface of the gravel as level and as flat as practical.

Constructing the Base. The base of the tank, its bottom, rests on the gravel. Its dimensions are exactly the same as the desired exterior dimen-

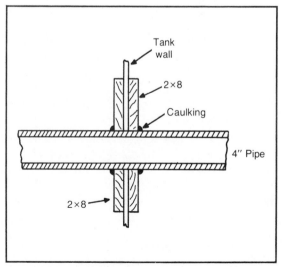

Fig. 18-9. How the wood walls of the tank can be reinforced to support the end of the sewer pipe.

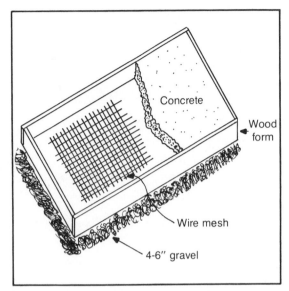

Fig. 18-10. Form and wire reinforcement that can be used to make poured concrete slab to support a concrete or brick tank.

sions of the tank. Bear in mind that the table giving tank dimensions for various gallons of capacity specify the inside dimensions of the tank. You have to allow for wall thickness.

Now you have to construct a wooden form into which you will pour the concrete that forms the base. If the tank is not going to be more than 4 feet wide and 6 feet long, you can make the base 4 inches thick. If the tank is going to be larger, it is best to go to a 6-inch base. When you go to the 6-inch slab, it is wise to reinforce the base with a section of 4 × 4 or 6 × 6 wire mesh, woven screen made for use in concrete. Wear gloves and use a bolt cutter to cut the screen. Cut the piece so that it is 4 inches shy of the sides of the form.

Use 2 × 4s for the 4-inch slab and 2 × 6s for the 6-inch slab. Simply nail them to form a box as shown in Fig. 18-10. Drive a few short stakes into the gravel and earth behind the form boards to hold them in place. Check to make certain the form is reasonably level. Now you are ready to pour the concrete.

Pouring the Bottom Slab. Compute the volume of concrete you need on the basis of the volume that will fill the form to its top. Typically, a 5-×-8-foot form that is 6 inches deep requires 20 cubic feet of concrete. Add 10 percent for waste and you still have less than a cubic yard. Mix in this small a quantity is very expensive. The mason years have a minimum charge. A quantity this small is also easy enough to mix by hand if you use a good-size mortar pan. In any case, use a 1:2¼:3 mix (1 part cement, 2¼ parts sand, and 3 parts crushed stone) because it is just about the strongest mix.

To compute the quantity you need, figure:

6 cubic feet of cement.

14 cubic feet of sand.

18 cubic feet of stone.

About 36 gallons of water.

To make 1 cubic yard, or 27 cubic feet of concrete.

In case you are wondering where everything goes, the sand fits into the voids between the stones, the cement fits into the voids between the grains of sand, and the water fills the balance of the openings. Note that a standard bag of cement contains just 1 cubic foot of cement. Use a shovel or a pail for measuring. You need not be any more accurate.

Do not attempt to mix the entire quantity of concrete you need at once. Do put a small quantity of the correct proportions of the materials into the pan and add water. In this way, you will not be mixing the entire weight at one time. Start by adding more water than you need. Mix until it is all one even, grey color. There is no need to mix beyond this point unless you need the exercise. Now add a little of the dry mix to bring the mix into the pan down to the right consistency.

You need to use about 6 to 7 gallons of water to each bag of cement in the pan. This will give you a sloppy mix that will find its own level. Add some Anti-Hydros to make the concrete watertight.

Dump the mix into the form. If you are going to use a screen reinforcement, spread the mix over the entire form to a depth of about 3 inches. Then place the screen on top and pour more concrete. No harm is done if anything up to a half hour passes between adding batches of concrete. A longer delay could cause the second layer of concrete to adhere poorly to the first.

Finishing. Fill the form with concrete. With a long straight board, held on edge, screed the surface of the concrete flush with the surface of the form. You will have to use a sawing motion. If there is a low spot in the concrete, simply add some more there and screed again.

Give the concrete an hour or two to set up and then spray it lightly with clean water. This will speed its *cure* (hardening). Wait three or four days before removing the form. You are now ready to lay the block.

Laying the Block. Use heavy-weight, 6-inch block. Do not use the lightweight or cinder blocks. You can use stretcher blocks (indented ends) alone or in combination with end block (flat ends). It doesn't matter because nothing will be seen.

Prepare the mortar by mixing *mortar cement* (cement that contains about 25 percent lime) with sand in the ratio of 1 part cement to 2 or 3 parts sand. Mix dry and then add sufficient water to bring the mortar to the consistency of ice cream. Mix no more than you expect to use in an hour or so. If you do not use the mortar in this time, discard it. If the mortar stiffens a bit during this period, add a little water and remix it. See Fig. 18-11.

With a mason's trowel, spread a bed of mortar 1 inch thick, 7 inches wide and 18 inches long. Start at any corner and spread the mortar along one edge of the slab. Set a block on this bed of mortar (small holes up, large holes down). Align the block with the corner and sides of the slab. Mortar consistency should be such that the block sinks a little into the mortar of its own weight. Make the block level by either judging it by eye or with the aid of a spirit level. If necessary tap the block down. If you have to raise the block, lift it off completely, add fresh mortar, and position it again.

Go to the second corner and repeat the operation with another block. Then do the other two corners in sequence.

Return to the first corner. Spread another layer of mortar on the slab so that it forms a right angle with the in-place block. Take a block and stand it on end. Butter its ends with mortar. With the trowel, press some mortar firmly against its ends. Now place this block against the side of the corner block with just enough pressure to form a ½-inch-thick layer (joint) of mortar between them. Make the second block level and flush with the edge of the slab. You now have a right angle formed by two blocks meeting at a corner. Do the same at each of the three remaining corners.

Using two bricks or two pieces of block as holders, stretch a line from one corner block to another. Position the line just clear of the front edges of the blocks. This line is your guide. Beneath this line, lay down a bed of mortar from one block to the other. Butter the ends of a number of blocks and position them along the edge of the slab—in line with the line.

If there isn't space for a full *closure block*, you have to cut one. Hold the block against the earth at an angle and repeatedly strike the block with the edge of your hammer along the desired break line. Do this to both sides of the block.

If the buttered mortar will not remain in place when you lower the block into position, place a stick behind the joint and force mortar into the joint with your trowel.

Repeat this operation until you have completed the first course (row) of blocks. Now start the second course at any corner. As before, start by laying down a bed of mortar an inch or so thick. Take care that the second course block rests on two blocks. In other words, the second block lies across the top of two blocks. Make certain the second block is vertical and flush with the lower block. Repeat at all the corners. Then fill in the spaces between the corners. Continue this way until you have gone as high as you prefer.

To accommodate the pipes, you simply leave space for them. Later you can position the pipes and lock them in place with mortar.

Finishing. After you have laid up block for a half hour or so, you have to point the joints. This is done by either running a pointing tool down the joints or using a piece of pipe or the point of your trowel. The purpose of pointing is to compress the mortar into the joints, making them stronger. At the same time use the side of your trowel to remove any excess mortar from the wall and to fill in any joints that you have missed.

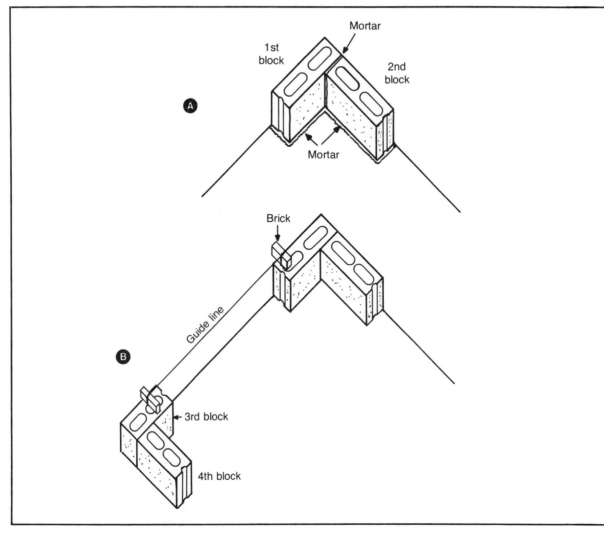

Fig. 18-11. Some steps in building a concrete block tank. A: Start at a corner. B: After two corners are laid, the space between is generally filled.

Dividers. Use 4-inch solids for the dividers. Simply lay them up as before starting with the concrete slab as the foundation. Because there will be very little pressure on the dividers, you need not make them integral with the wall. You can simply mortar them in place.

Plastering. The interior of the tank is plastered with mortar cement to increase its life and make certain it is watertight. Make the mortar plaster from the same mortar cement (sand mix-

tures previously described), but add a little Anti-Hydros to the mix. This will speed setting and make the plaster more watertight. See Fig. 18-12.

Use a steel trowel. Apply the mortar to the slab a few inches in from any wall. With a sweeping motion, spread the mortar up the side of the wall. Try to hold it to a thickness of ⅜ of an inch. When this has set up (hardened a bit), roughen its surface with a comb made from a dozen nails driven into a board. Give the first layer of mortar a few days to

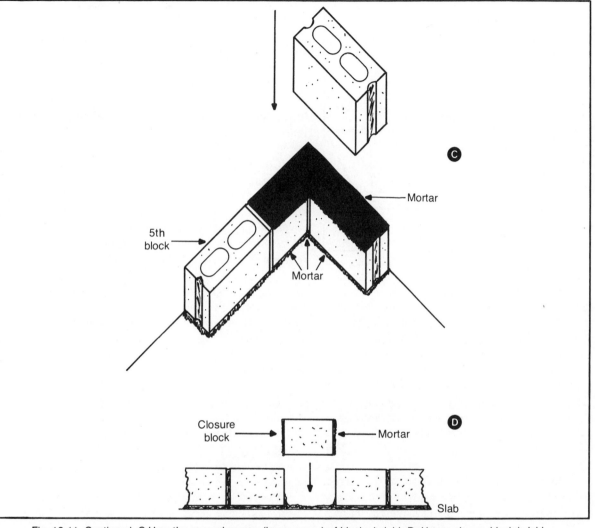

Fig. 18-11. Continued. C:How the second course (layer or row) of blocks is laid. D: How a closure block is laid.

harden and then apply a second ⅜-inch layer.

The floor or bottom of the tank should be treated in the same way at the same time. You can't step on the wet plaster without ruining it, but you can lay down a large, flat board and stand on that while you work.

MAKING THE ROOF OR TOP

You can construct the roof of reinforced, precast slabs (available at mason supply yards), you can use 3-inch bluestone slabs, or you can pour the roof of concrete, either in place or in a nearby form and then slide it over. It doesn't have to be a single slab of concrete; you can make several. The choice depends upon tank size, availability, and cost of stone or precast slabs.

Before you position the masonry cover, no matter what material you use, you have to make

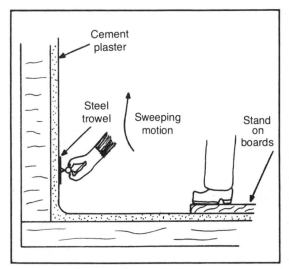

Fig. 18-12. The inside of the block (or brick) tank should be cement plastered to extend its life.

certain that the top of the tanks four sides are level and equal in height. You can do this with a long, straight board and a spirit level. If you find one spot low, build it up with mortar. To keep the mortar from simply dropping into the holes in the block, stuff each hole with wet newspaper. It is most important that the top surface of the tank be on one plane because none of the masonry covers you use will bend to conform to irregularities. You must have a watertight and airtight seal here. See Fig. 18-13.

Precast Slabs. When they are of the correct size and assembled, precast slabs do not have to cover the tank exactly, but may overlap without problem. Precast slabs present only a single difficulty and that is cutting the port hole.

Precast slabs are made for use as steps, porches, and similar uses. They do not come with large holes that you can use for a clean-out port. You have either to cut one with a chisel—and this is hard, time-consuming work—or you can let one slab remain in place by weight alone. To get to the tank's insides, you have to remove a lot of soil and slide the slab back or raise it.

In any case, you apply an inch of mortar to the top of the tank to hold the slabs you want to remain immobile. Then add mortar atop the joints between the slabs to seal the cracks.

Casting Your Own Slabs. Construct one or more forms from 2 × 4s nailed on edge to a sheet of plywood. Make the inner dimensions of the form such that you need an even number of cover slabs to cover the tank. Oil the inside of the form with motor oil (new or used) so that you will have no difficulty removing the casting from the form.

Use a 1:2¼:3 mix with small crushed stones, a little on the sloppy side. Mix thoroughly and cover the bottom of the form with about 1 inch of the mixture. Then position the steel reinforcing bars shown in Table 18-1. Cover with more concrete and screed. The latter action consists of "sawing" a straightedge board across the filled form to make the concrete level with the sides of the form.

Give the concrete two days to harden a bit, and then lift it carefully out of the form and place it on a flat section of the earth. Sprinkle it lightly with water to speed curing. In a week, it will have reached 85 percent of its full strength and you can position it. See Fig. 18-14.

Providing a Port Hole. When you cast your own roof slabs, you can incorporate a short section of a large-diameter piece of vitrified clay pipe in one or more of the slabs (Fig. 18-15) or you can leave an opening in two adjoining slabs and provide an

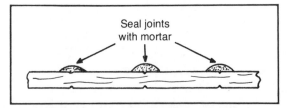

Fig. 18-13. When individual precast slabs are used to make a tank top, the joints have to be sealed with mortar.

Table 18-1. Reinforcement Used with Cover Slabs.

Span*	Bar size	Bar Spacing	Crossbar Spacing	Crossbar Size
4 ′	⅜′	10 ′ ″	18 ″	⅜ ″
5 ′	⅜ ′	8 ′ ″	18 ″	⅜ ″
6 ′	⅜ ′	6 ′ ″	18 ″	⅜ ″
8 ′	½ ′	7 ′ ″	12 ″	⅜ ″

*Slab to be a nominal 4 in. thick.

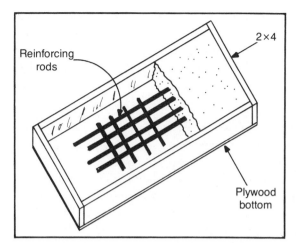

Fig. 18-14. How 2 × 4s can be used to pour a tank-top cover. Reinforcing rods are required.

opening this way. Use 6-inch pipe because you have to poke a hose end into the tank and sometimes you will also have to get a shovel down there.

When you make the port hole out of pipe, you want to include a hole every few feet and at least one above each section if the tank has more than one section. If you make a trap door opening, one will do if you can reach all the sections from the single opening. The door can consist of a slab of 3-inch flag or bluestone, held in place by weight, or a small cast slab.

Monoslab Roof. There is still another way to roof the masonry septic tank. You can pour its roof in a single pour. This is the best method because it produces a solid, cemented-in-place slab that helps strengthen the tank.

The technique consists of building a wooden false floor inside the tank and then surrounding it with a wall of wood. The concrete is poured atop the floor; the wall of wood serves as the form.

Use dimensions shown in Table 18-1 for slab thickness and necessary reinforcing rods.

To support the false floor, construct a frame of 2 × 4s that fits inside the tank. The frame is supported on more 2 × 4s. In addition, install crosspieces every 12 inches or so, running horizontally from one side of the frame to the other. These crosspieces are supported by more 2 × 4s, every 2 feet or so. Bear in mind that you will be

holding up some 3000 pounds. That is about what a yard of concrete weighs. See Fig. 18-16.

To provide the port hole, cut or leave a 2- × 2-foot hole in the false floor and "box" it with 2 × 4s on edge.

Cover the floor of the form with a couple of inches of concrete. Position the steel bars and then pour the balance of the concrete. As before, screed its surface level. Give the concrete a week to harden properly; keep its surface moist during this period. Now you can enter the tank and remove the underside of the form and support timbers.

Constructing the Form for the Walls. If the earth in the hole is hard and firm, and the hole is not too deep, you can possibly construct a single-wall form and pour the concrete in between the form and the soil.

If, as is often the case, the soil is too soft, you have to construct a double-wall form. Use 2 × 4s for the verticals, backed by pairs of 2 × 4s, called walers. Their number and spacing depends upon the height of the form. Bear in mind that it is the height of the form that determines the outward pressure on the form. Wet concrete acts like water and the greatest pressure develops at the bottom of the form. It can amount to tons.

Because concrete sets up fairly quickly, you can reduce this force by pouring the concrete very slowly into the form. You should not stop pouring until the form is completely filled and its top has been screeded flush with the form. The pour has to be one homogeneous mass.

To make certain the concrete has reached all

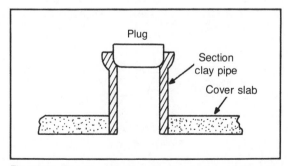

Fig. 18-15. One way of making a port in a tank cover. The clay section must be large enough to accommodate a long-handled shovel.

203

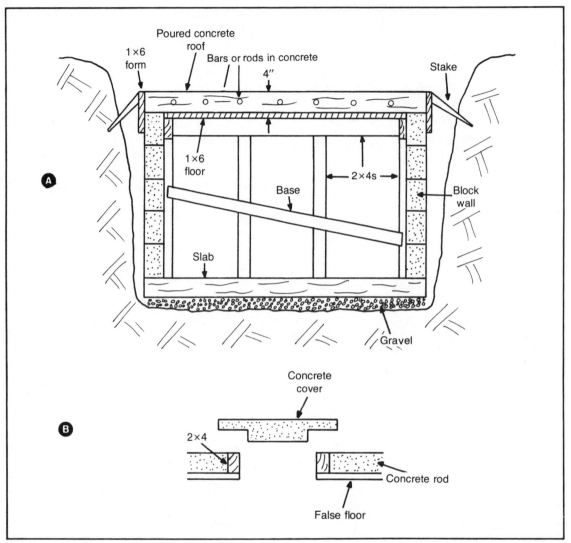

Fig. 18-16. A: How a slab can be poured atop a block or brick tank. The wood form is removed after the slab has hardened. Access is through the port hole. B: How an aperture may be left in the poured slab to accommodate the concrete porthole cover.

corners, it is advisable to poke a stick into the concrete in the form to help it settle. Jarring and vibrating the concrete makes it a little stronger and more compact.

Use either ⅜-inch plywood or 1-inch roofers to line the form. If there are holes and openings, patch them from the outside of the form.

Compute the concrete on the basis of the volume needed for the form, plus 10 percent for waste.

If you require more than a couple of yards and have no helper, you will probably be well advised to use Ready-Mix. It will be a lot easier and may very likely be considerably less expensive.

Give the concrete three or four days to cure a bit and then carefully remove the form. If you wait too long, the board will stick to the concrete. If you work too quickly, the still-soft concrete might break. With this done, carefully back fill to further protect the concrete tank. Wait a few days longer and then construct the roof or cover. Use any of the methods previously described.

Plastering the Concrete Tank. The same two layers of mortar cement previously described for use with the block and brick tank can be applied to the inside (but not roof) of the concrete tank. Give the plaster three or four days to dry hard.

ASPHALTING THE INSIDE OF THE TANK

The inside of wooden tanks and masonry tanks must be coated with asphalt. Use a steel trowel to spread the asphalt on the ceiling, the walls, and then the floor. To keep from stepping on fresh tar, spread the tar all around yourself. Then lower a ladder into the tank and stand on a rung to finish the work. If you have built a tank with pipe sections for the port holes, you have to asphalt the tank before you cover it up.

Protection. With your tank completed and covered, it is a good practice to plug the inlet and outlet ports with wet newspaper or cap them with bags to keep debris from accidentally entering.

MAKING A BRICK SEPTIC TANK

Start by pouring the slab as previously suggested. Lay down a ½-inch thick band of mortar, 5 inches wide and 10 inches long. Start at any corner of the slab and follow any side. Bear in mind the instructions given for use with concrete block and do the same with a brick. Align it with the corner of the slab and the edges forming that corner.

A standard (common) brick is 2¼ × 3¾ × 8 inches in size. Lay it on its flat side. When it is in place, butter the edge of a second brick and position it alongside the first. Be careful that you do not disturb the first brick as you do so. Then do the same at the other corners.

Next, complete each corner by laying a pair of bricks at right angles to the first pair at each corner. Again you want ½ inch of mortar beneath the brick and an equal thickness of mortar between the bricks. Follow the procedure with the line to guide

you in positioning the bricks between the corners.

Return to the corners and lay more bricks there. Take care to position the upper bricks over the joints between the lower bricks. Make certain the bricks are flush and vertical. The basic procedure is exactly like that used with concrete block. The bricks are easier to handle, but there are lots more bricks and you have to make the walls two bricks thick. Make the outer walls first, and then the inner walls. This is the beginner's way. Just see that the outside vertical joints are not in line with the inner joints, and spread mortar between the walls.

Give the brick three or four days to set up; a little moisture will help. Then plaster the inside with mortar plaster as previously described and construct the cover or roof as described.

To accommodate the pipe, simply leave openings in the brick, slip the pipe through, and cement it in place. The dividers can be made of brick or they can be made of solid concrete block set on edge.

If the tank is going to be less than 4 feet high, you can build it with SCR brick, which is 2⅛ × 5½ × 11½ inches in size. The roof for the brick tank can be constructed exactly the same as any of the other masonry covers previously described.

THE POURED-CONCRETE TANK

The poured-concrete tank is the best of the bunch. Made properly, it will probably last the longest. Start by preparing a slab foundation as previously described, with two reservations.

One, the walls should be poured within a few days of pouring the foundation. If you don't, the portions of the foundation that will carry the walls should be coated with Will-Bond. Otherwise, the joint between the new and the old concrete will or may be weak.

Two, a little Anti-Hydros should be added to the mix used for the walls. This will help keep the concrete watertight. You have, of course, already added some to the concrete you used for the slab. Mason supply yards carry this product or a similar mixture.

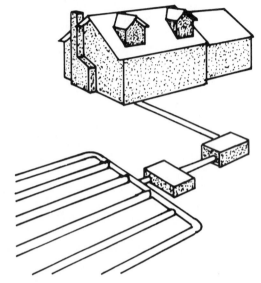

Chapter 19

Running the Pipes

I N THE PRECEDING CHAPTERS WE DISCUSSED THE angle or pitch at which the soil pipe carrying the black water should be run. The ideal pitch has been found to be ¼ inch to the foot. In practice, this is usually changed to ½ inch to the foot. When and where a greater pitch is required, the sewer line must be run at a 45-degree angle or more. Otherwise the muck will hang up along the pipe run.

PIPE TYPES AND DIMENSIONS

While the building codes throughout this country are not carbon copies of one another, they are still fairly close in many details. One such detail is the diameter of the house sewer pipe. The codes usually call for a 4-inch pipe (meaning a pipe having a 4-inch internal diameter). Because copper and plastic pipes are much smoother on the inside than cast iron (the old-time sewer pipe), most codes permit 3-inch or 3½-inch pipe to be used as sewer pipe. It is a mistake to run anything smaller than 4-inch sewer pipe even though the pipe inside the house is only 3-inch or 3½-inch pipe. Unless you

install a clean-out in the sewer line inside the building, it is very difficult to clean this section of the line. Going to the larger pipe diameter reduces the possibility of blockage. See Fig. 19-1.

Transition Fittings. Transition fittings are used to connect pipe made of one type of material to a pipe made of another type of material. Examples are cast-iron pipe connected to plastic pipe or copper pipe.

Reducers. Reducers are fittings designed to be used with one type of pipe when it is to be joined to a pipe of a different diameter, but of the same material. For example, copper-to-copper or cast iron-to-cast iron.

Choice of Pipe Materials. The best material is still cast iron. It is the strongest and it will probably outlast the others. In some homes, cast-iron sewer pipes have been in continuous use for a hundred years and more.

Always use cast-iron pipe when the pipe goes beneath the concrete cellar floor of the building—at least in the under-concrete section. Always use

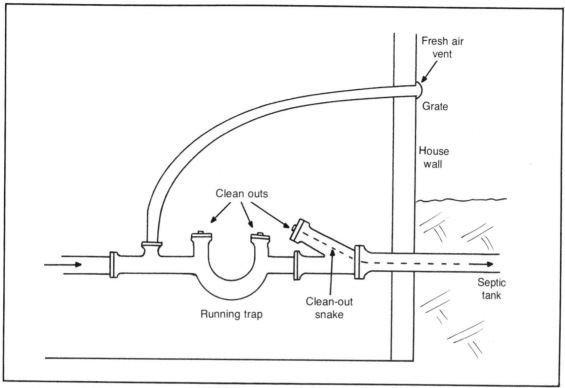

Fig. 19-1. In a modern home sewer connection, there is always a fresh-air pipe plus a running trap followed by a clean out. The stack, the vertical pipe running from the house sewer line through the roof, is usually to the inside of the fresh-air connection.

cast-iron pipe where the pipe has to pass under a walk or a driveway or road.

Never install copper pipe beneath concrete or beneath the earth. If you are going from copper to cast iron by means of transition fittings, always make certain the fitting is sufficiently clear of the earth or concrete to protect the fitting from moisture. If the fitting is covered with a film of moisture, galvanic action will quickly corrode the pipes at the joint.

Fitting Dimensions. Fittings are defined by their dimensions. Thus a 4-inch elbow is made to accept two 4-inch pipes. With cast iron, copper, and clay, the size given always refers to the pipe's internal diameter. In the case of plastic pipe, the outside diameter is the figure usually given. To be certain, always try the plastic fitting on the plastic pipe. It should make an interference fit (it should be snug). If the fit is too tight, you will not be able to

get sufficient cement into the joint. Too loose a fit and you are asking the cement to span a larger gap than for which it was designed.

Choice of Fittings. Do not use bends and angles, but do use long curves. The difference is that the bend or angle makes a sharp turn. The long curves make the same angle but takes longer to do so. The first type of fitting restricts the flow of liquid more than the second type does.

Transite and Orangeburg. The transite pipe is made from concrete and asbestos. It is fireproof, rotproof, and bugproof. If it not as strong as copper or cast iron, but in a way it is stronger than copper. It will carry a greater weight, but it is brittle and will shatter under load (while copper will bend). Orangeburg is the trade name from pipe made from mineralized fiber. It is the least expensive of all the pipe used for sewage and storm water, but its life is limited because the tar with

207

which it is impregnated eventually dissolves under the attack of earthy microbes and acids.

TANK EFFLUENT PIPING

There has to be some pitch to the pipe leaving the septic tank, but its pitch or angle is not at all crucial. You can make it anything convenient because the effluent at this point is a liquid. There is and there should be no solids in it. Thus it can flow down any angle at all without problem.

Material and Size. The size of the pipe leading effluent from the septic tank to the drain field should have the same diameter as the pipe leading to the tank. The pipe leaving the tank should have the same diameter, but there is nothing to be gained by going to the larger, more expensive pipe.

Any of the piping materials mentioned, with the exception of copper, can be used here. Generally, vitrified clay is the pipe material selected.

DRAIN FIELD PIPING

The drain field pipe performs two functions. It guides the effluent through the drain field and at the same time the pipe permits the liquid to seep into the adjoining earth.

In years past, the drain field piping was usually made of short sections of fired clay, generally octagonal in their exterior cross section and circular internally. Called *drain tile* they are spaced about ⅛ of an inch apart so that the liquid can filter into the earth through the spaces provided. Today, plastic drain pipe is more often used. The plastic pipes come perforated, and it is easily and quickly joined. The perforation are positioned sideways.

RUNNING SEWER PIPE

The term *running* in plumbing parlance means to cut, fit, and install pipe. The methods or techniques used to run pipe varies with the pipe. Cast-iron and vitrified-clay pipe are manufactured with bells and spigots. The bell is the enlarged end and it is designed to accommodate the spigot of a following section of pipe. The other piping mentioned is usually joined by what may be termed a *slip fitting*.

Cutting Cast Iron. There are three ways to cut cast-iron pipe. One is to score the line with a hacksaw—going around and around, deeper and deeper—until you have cut a groove about ¼ inch deep. Then a light tap with a hammer should cause the pipe to come apart on the line.

A second method consists of striking the line of break or cut with a dull cold chisel and a hammer. Keep pounding lightly on the line, going around and around the pipe, until it breaks along the line. To do this properly, you have to support the pipe on a block of wood or a pile of earth. The point of the chisel compresses the fibers of the steel until the pipe comes apart itself.

The third method is the easiest, but it requires a pipe cracker. This is sort of a giant pair of pliers that is used to apply pressure to the pipe, and the pipe breaks along the preferred line—most of the time. You might be able to rent this tool locally. Perhaps your local plumbing supply house has one you can borrow. Cutting cast iron pipe with the "cracker" takes but a minute.

Joining Cast-Iron Pipe. The standard or usual method consists of setting the spigot of one pipe within the bell of another. Both pipes must be perfectly aligned. With the aid of a caulking chisel, a layer of *oakum* is forced into the space between the bell or hub of one pipe and the spigot of the other. Oakum is a soft, coarse rope made for this purpose. It is wound around the pipe before being forced into the opening. You want sufficient oakum to fill about ½ inch of the opening. In place of a caulking chisel, you can use a short piece of hardwood. The oakum seals the joint.

Next you fill the remaining space between the spigot and the hub with lead. You can use hot lead that is melted in a pot and poured in or you can use *lead wool*, which is lead cut into fine wires and pound it in. See Fig. 19-2.

In either case, you want at least 1 inch of lead but not so much you fill the opening. Next the lead must be "spread." This consists of striking it with a dull chisel. The blows expand the lead and lock the two pieces of pipe together. If the joint is not disturbed, it will last as long as the pipe (which is a long, long time). See Fig. 19-3.

An alternative method consists of using No-

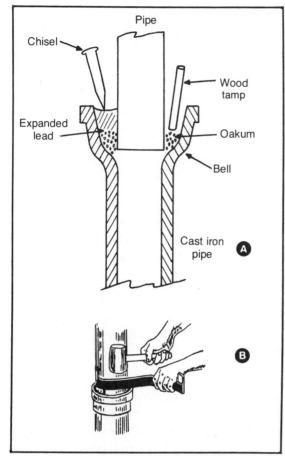

Fig. 19-2. A and B show the steps in caulking a cast-iron pipe joint. The lead can be poured (or in wool form, fine wire) and pounded in order to expand it. For a temporary joint or an iron-to-clay joint, mortar can be used in place of the lead. If this is done, it is brought back farther along the spigot.

Hub fittings. To use these fittings, you must first remove the bell or spigot. In this way you end up with "straight" pipe (pipe without a bell or hub). The No-Hub fitting consists of a short section of plastic tubing and two stainless steel clamps. The tubing is forced over the pipe ends. The clamps are used to hold the tubing in place. The advantage of the No-Hub joint is that it is very fast (when you have a pipe cracker to remove the hubs); the joint is slightly flexible and will sustain light shocks; and the joint will bend a few degrees, something that is impossible with a lead joint or the cement joint.

Still another way of joining sections of cast-iron pipe consists of using oakum followed by cement (the kind used for masonry work.) The spigot is positioned within the hub and the oakum is pounded home. This is followed by a layer of mortar cement made by mixing regular cement and clean sand in the ratio of 1 part cement to 2 parts sand. This type of joint is deemed the least desirable of all types of cast-iron pipe joints. Its use is generally confined to pipe that will be positioned within the earth.

To run cast-iron pipe a specific distance, that distance must either be an exact multiple of pipe length or you have to cut some pipe to make it fit. The problem that might come up is that you could end up requiring a very short section of pipe to complete the run. Ideally, you would start the

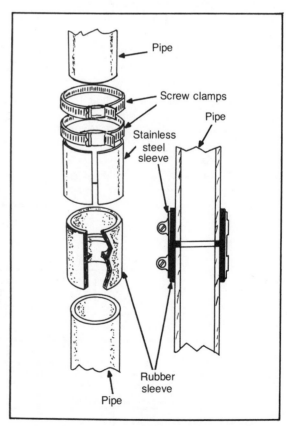

Fig. 19-3. An exploded view of a No-Hub joint. You can either purchase cast-iron pipe with a hub or cut the hub off.

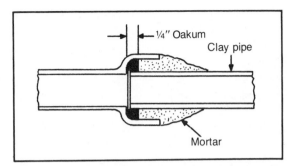

Fig. 19-4. A clay pipe joint. Note that the oakum must be about ½-inch thick and that plenty of mortar cement must follow. This is a bell and spigot joint.

sewer pipe run at the septic-tank spigot on the inside and then add pipe sections until you reached the house and worked your way up the soil stack to the roof. But this is generally impractical. Generally, the house plumbing is completed first and the pipe is run from the house to the septic tank.

The solution is to run the last four or five sections "dry." This will enable you to see exactly how much pipe you will need. Then you can, if needed, cut a little from several lengths of pipe and thus not end up requiring a 3-inch or 4-inch section that is just about impossible with cast iron. Of course, you can also use some No-Hub fittings. See Fig. 19-4 and Table 19-1.

Another problem with cast iron is that you have to end up with a hub facing the tank so that you can use a standard fitting inside the tank. This can be accomplished by running the sewer line all the way to the tank. The last fitting on the line, the fitting that will be inside the tank, is slipped through the hole in the tank wall—spigot first—and into the bell on the last section of sewer pipe.

This is a problem you will only have with cast iron and vitrified-clay pipe. It is no problem with plastic pipe because you can easily shorten plastic pipe as little as ¼ inch and join it to the following section with a slip fitting. A *slip fitting* can be slipped up any distance along a pipe and then slipped down to cover and make the joint.

Cutting Vitrified-Clay Pipe. The pipe is raised either on a board or a pile of earth so that its hub is clear of the ground. Then with a cold chisel and a hammer, a line is scored all around the pipe along the desired line of cut. This is repeated again and again, and the pipe is turned. Eventually, the pipe will break apart along the line. The trick to the technique is patience.

Joining Vitrified-Clay Pipe. Joints are made with oakum, followed by mortar, as previously described. The joints are weak. The joined pipe can be positioned vertically without many sup-

Table 19-1. Clay and Cast-Iron Pipe Dimensions.

				Cast-Iron Soil Pipe		
					Weight (lbs.) per 5' length	
Size	A	M	J	T	Single Hub	Double Hub
3	4.19	3.88	3.50	0.18	45	47
4	5.19	4.88	4.50	0.18	60	63
5	6.19	5.88	5.50	0.18	75	78

		Vitrified Clay Pipe			
Internal Diameter Inches	Laying Length Feet	Inside Diameter at Mouth of Socket Inches	Depth of Socket Inches	Minimum Taper of Socket	Thickness of Barrel Inches
4	2	6	1½	1:20	9/16
6	2	8¼	2	1:20	⅝
8	2, 2½, 3	10¾	2¼	1:20	¾

ports. If the pipe is laid horizontally, it must be supported. Beneath the ground, the earth provides the support. Above ground, as in a cellar, a concrete block or similar support should be provided at every joint.

All the problems of pipe size and fittings relating to cast-iron pipe will be encountered with clay pipe. Note that clay pipe can make direct contact with iron pipe without deterioration. When an iron pipe enters a clay bell, the joint is made with oakum and mortar. When a clay pipe is placed within a cast-iron bell, the joint can be made with oakum and hot lead. But the lead must be expanded very gently.

Cutting Plastic Pipe. Plastic pipe can be cut in the large diameters with any type of wood-cutting saw. The fine-toothed crosscut saw probably works best. Wrap a straightedged strip of paper around the pipe. That will give you a straight line to follow with your saw. Small-diameter plastic pipe can be cut with a tubing cutter (the kind used for copper tubing). Once you have cut the pipe, remove any burrs that might have formed with the help of a file or sandpaper.

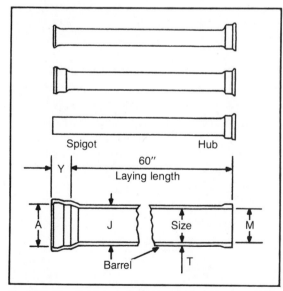

Fig. 19-5. Top: Typical, standard forms of cast-iron pipe. The No-Hub pipe is usually a special order. Bottom: The major dimensions that are often specified when ordering cast-iron pipe. Laying lengths vary.

Because plastic pipe can be cut to any length, you will face none of the dimensional problems that will be encountered with both clay and cast-iron pipe.

Joining Plastic Pipe. The actual joining of one section of plastic pipe to another or to a fitting takes less time than it takes to describe the operation; it is that fast.

To join two pipe ends, use a coupling of the correct size. The shine normal to plastic pipe is removed with steel wool, sandpaper, or a file for a distance equal approximately to the diameter of the pipe. This is done to both pipe ends to be joined. You can also use a precement solution, but it usually is not worth the trouble.

Let us assume you are joining two lengths of pipe by means of a coupling or a slip fitting. A coupling has an inner lip that prevents the fitting from being slid the entire length of the pipe. A slip fitting does not have a coupling; it has no internal lip. In any case, the rotation of one pipe in relation to the other is unimportant.

A band of plastic pipe cement is applied to one pipe end. The shine has been removed from this end. The band of cement should be just as wide as the diameter of the pipe and should be applied in one continuous sweep all around the pipe. The fitting is slipped into place, given a quarter turn, and left untouched for a minute or two. Then the other pipe is treated in the same way.

Now assume that you are going to fasten an elbow on the end of a plastic pipe. The rotation of the elbow in relation to the pipe is important. Assemble the joint dry. With a pencil, draw a line on the pipe and the fitting. The line on one runs into the line on the other. Disassemble the joint and apply the cement as before. Reassemble the joint. Give it a fraction of a turn, ending up with the two pencil marks in a line. Thus, the orientation of the parts is maintained. Let the joint be for a minute or two.

You have applied sufficient cement when a bead of cement appears all around the edges of the fitting. More is simply a waste. Use the correct cement for the particular pipe material you are using (PVC cement for PVC pipe, etc.). The cement must be at room temperature (place it in a pan of hot water if it is cold). If the cement has become "ropy" and thick, thin it with cement solvent or discard it.

Now for the bad part. Once made, a cemented plastic joint cannot be taken apart. The parts are "welded" together. If the joint leaks, cut the section out and replace with a short section and two slip joints.

Cutting Transite. Transite can be cut with a hacksaw. It is normally available in 5-foot and 10-foot lengths.

Joints and Fittings. The pipe is manufactured with tapered ends that are designed to be force fitted into couplings. The couplings have internal rubber rings that seal the joint. No problem. Problems arise when you have to join the cut ends of Transite. You can cut the necessary shoulder into the pipe end with a file or power grind stone or you can try to secure a No-Hub fitting that can be adapted to the job.

Masonry yards and the like rarely carry more than a few basic Transite fittings. If you are going to require more than a T or a long radius bend, it is best to order the fittings well in advance of actual construction.

Cutting Bituminous Fiber Pipe. Bituminous fiber pipe is manufactured in lengths of 8 feet and 10 feet as standard sewer pipe and as drain pipe. The latter type has one or two rows of holes for drainage. The pipe is light in weight, easily cut with a carpenter's saw, but is not acceptable everywhere as a sewer pipe. Note that it comes in a number of grades.

Joints and Fittings. Like Transite, bituminous pipe fittings come in a very limited number of types and dimensions. If you have anything

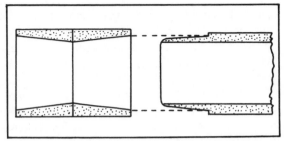

Fig. 19-6. The cross section of a bituminous fiber pipe and coupling. This is a force fit. No gasket or fittings are used.

beyond the usual right turn or 45-degree bend in mind, you had best make certain they are manufactured before you begin your project.

Like the Transite fittings, the bituminous pipe fittings are pressed into place with the aid of a block of wood and a hammer. The bituminous fitting, however, has no internal gaskets. The joint is made by the tight fit itself. Like the cut section of Transite, when you cut bituminous pipe you have to file or grind a shoulder on it to get the fitting to go on. This has to be done very carefully because there is no internal, flexible gasket to fill the voids. See Fig. 19-6.

Bituminous pipe is the least desirable of all the piping material. Bituminous pipe no doubt has the shortest life span of them all.

RUNNING THE DRAIN-FIELD PIPING

Because the effluent leaving the septic tank will be almost pure liquid, you can run this piping at any convenient angle that is more than ¼ of an inch to the foot. Everything suggested for the piping from the building to the tank pertains to the piping that will be installed from the tank to the drain field.

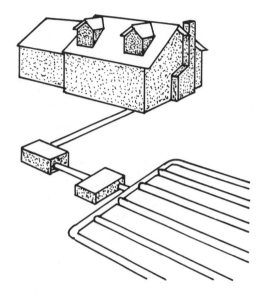

Chapter 20

The Percolation Test

THE DRAIN FIELD OR THE LEACHING FIELD, AS some people call it, is the key to local sewage treatment. You can position the septic tank almost anywhere. So long as the building is higher than the tank and the tank is higher than the drain field, the system will work. The drain field *must* drain. That is the controlling factor in the entire operation.

LOCATING THE FIELD

The basic rules for locating the drain field (Fig. 20-1) are that:

☐ The field should be as far away from all wells (yours and your neighbors') as possible.

☐ The field should be on ground lower than the ground on which the well(s) is situated.

☐ The spot selected should not be a natural sump where rain collects and remains.

☐ The field should be positioned on the sunny side of the building or property.

☐ The field should not be close to a tree as tree roots can clog and destroy the drain-field pipes (tree roots reach as far as the tree's branches).

Trees throw heavy shadows.

☐ The earth in which the drain pipes will be placed should be highly permeable.

Highly Permeable Earth. The ideal drain field soil is light, airy, porous and dry. This is nice to know, but what you must know before you proceed to even think of a septic system, tank, and field is what sort of soil do you actually have? If you have any choice in locating your drain field, where does the best drain-field soil exist on your property?

Neither of these questions can be answered by just looking at the surface of the ground. To be reasonably certain of what you have, you need to dig a number of test holes. Each hole should be at least 2 feet deep. You cannot depend upon what one or two holes may disclose.

The earth is often naturally stratified. If you are on the bed of an ancient river, for example, going a few feet in any direction might uncover different soil; soil that was washed up onto a bank by the now forgotten river. If there has been extensive building in your area, you might not uncover

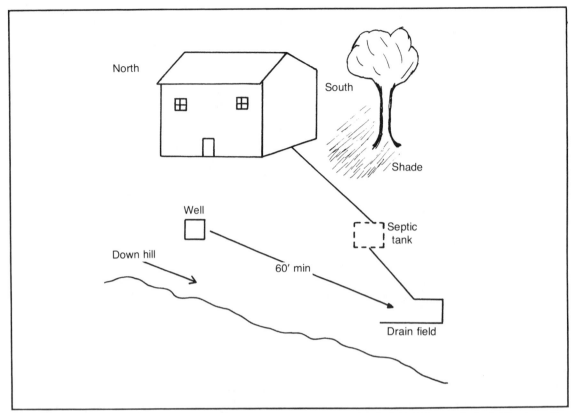

Fig. 20-1. Take a last look at the area you have chosen for your septic tank and drain field. Make certain the field is clear of the well, clear of the shady side of the house and shade tree, and on the downhill side of the property.

the virgin soil. You might be digging into something the builder dumped. A single hole will not reveal the nature of the soil throughout the entire area.

Another factor to consider is depth. The soil near the surface might be heavy with undesirable clay, but there could be highly desirable sand just below. On the other hand conditions might be reversed. The upper layer or horizon might be sanded with solid clay immediately below. You have got to be reasonably certain, and that means digging many test holes to a depth a foot or more below planned trench depth.

What to Look For. The best soil is pure sand and the absolute worst is no soil at all—solid rock. Solid clay is not much farther behind in the unwanted list. What about in between soils? How do you determine where a particular soil falls between the two extremes?

If you have a soil conservation office in your area, you could bring a sample to their office and have its composition tested. Your local building department or health department should be able to direct you to the nearest SCO.. Take care to sample the earth at the depth you plan to position your drain pipe. Make certain your sample is representative of the entire area and not a spot of sand the building contractor dumped.

You can make a crude estimate of soil quality by its feel. Pure clay can be worked almost into the form of a toothpick and it will hold its shape. Sand, even when soaking wet, can hardly hold a ball shape without collapsing.

Try to shape a damp sample of the soil in your hands. The thinner you can work or shape it the greater the percentage of clay that is present. When the soil will only hold a large, crude shape, you know that there is considerable sand present. The degree of plasticity exhibited by the soil sample between these two extremes indicates its clay/sand ratio.

There are exceptions. If the soil is rich in humus, which is decayed organic material, the soil will not be particularly loose, yet it will be low in clay content. Humus is generally limited to the top few inches (the topsoil). It is dark in color and so is easily recognized. Subsoil is usually light in color, and your drain field will go into the subsoil.

Another exception may occur when the soil is slopping wet. In this condition, it will exhibit much greater plasticity than when merely moist.

Another test consists of placing a ball of the soil in a pan of water. Pure clay will just lie there. Puddling the water will eventually cause the clay to dissolve. Pure clay thoroughly mixed with as much water as it will absorb will form a slippy, shiny mud. Clay and sand will tend to separate. Note that pure clay can be formed into a cup that will carry water safely. The cup does not have to be fired.

PERCOLATION TEST

The purpose of the percolation test is to learn by actual experiment the rate at which the soil in your proposed drain field can absorb water. This is done by pouring water into a hole in the ground and timing the rate at which the water disappears.

Digging the Holes. If the plan is to run the drain pipe (Fig. 20-2) in more or less a straight line along fairly level earth, you need at least five test holes. If the drain pipe is scheduled to be laid in more or less of a spiral, you require at a minimum three test holes. If there is to be one main-feed drain pipe separating into several branches so that overall the drain field is confined to a roughly square or rectangular area, three test holes should do it.

The reason for three or more test holes is simply that you want to be certain that you are actually determining the percolation rate of the entire drain field and are not by chance making your test in a pocket of sand.

Hole Depth. The depth at which you place the drain pipe depends upon climate. The warmer the

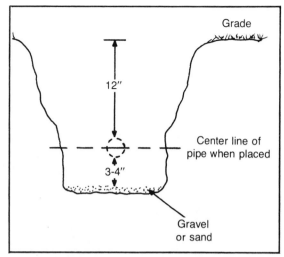

Fig. 20-2. The top of the perforated pipe and the pipe leading to the field should be about 12 inches at a minimum below the surface for protection from traffic and light frost.

climate the closer you can place the drain pipe to the surface of the soil. If there is any chance that anything heavier than people will pass over the field, the drain pipe should be 1 foot or more below the surface. This distance is measured from the top of the drain field pipe to the surface. Where there is frost, the drain pipe must be positioned sufficiently deep within the soil to preclude freezing of the effluent. This dimension poses a problem. Generally, the local frost line depth as stated by the local building code includes a very generous margin of safety. Usually this figure will lead you astray. Not only do you not need the margin of safety, the effluent flowing into the drain pipes is not pure water. Therefore its freezing point is considerably below 32° Fahrenheit. What figure should you therefore use for drain field depth? The depth most often quoted for this purpose is between 18 and 24 inches. Check with your local building department to learn their recommended drain field depth or at least what they believe is suitable for their community. Check with your neighbors to learn their drain field depth and their winter experiences with their field.

Aerobic bacteria work best when they have maximum air and, naturally, the deeper you plant the drain pipe the less air the little creatures will

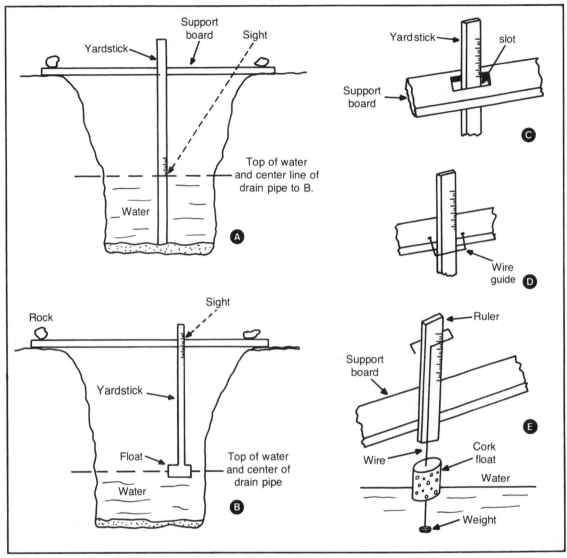

Fig. 20-3. Some ways to measure the fall of test water poured into a test hole. Note that the top line of the test water should be at least level with the planned position of the center of the drain pipe. The walls of the holes should be reasonably vertical at and below this depth. A: A yardstick placed in the water and rested against a horizontal board can be visually sighted at the water line. B: The yardstick can be mounted vertically on a wood float. C: A slot through the support board will hold the yardstick in a vertical position. D: A piece of stiff wire can be bent and fastened to act as a guide for the yardstick. E: Here a cork is used as a float, a length of stiff wire indicates the height of the float on the ruler nailed to the support board.

have. Thus you want to dig vertical-sided holes, with relatively flat bottoms, to a depth of 3 to 4 inches below the planned resting place of the bottom of the drain pipe. Just cut as straight down as you can with your shovel. Do not pat the sides of the hole with your shovel because doing so will com-

pact a little of the earth on the face or walls of the hole, altering your test results. If the upper portion of the hole breaks away and falls to the bottom, just scoop it up. You just need the lower sides of the hole to be reasonably flat and vertical. This is where you will pour the water.

Conducting the Test. Fill the bottom of the hole with a few inches of coarse sand or gravel. Make the surface of the gravel flat. Fill the hole with sufficient clean water to fill the hole to a level approximately parallel to where the center line of the drain pipe will rest when it is installed. Now measure the rate at which the water is absorbed.

Measuring the Rate. The rate can be measured any of several ways. No doubt you can easily devise additional methods.

Place a flat stick across the test hole. Keep the stick in place with a pair of stones. Position a yardstick vertically in the hole and rest it against the stick. Then carefully pour water into the hole until it is up to the desired level. Read this level on the yardstick. Time the rate at which the water level drops against the markings on the yardstick.

Fasten a ruler or a yardstick (Fig. 20-3) more or less the center of a block of wood, in a vertical position. Fill the hole to the desired level with water.

Place the yardstick and the attached block, which now functions as a float, on the surface of the water. Read the starting depth of the water by noting the position of the aforementioned horizontal stick against the markings on the yardstick. Time the rate at which the water level drops against markings on the yardstick.

Neither of the two devices are much trouble if the flow is fairly rapid. If it is slow, it is tedious work guiding the yardstick in position for an hour or so. One alternative is to cut a square slot in the stick that is placed across the hole. The yardstick is slipped through the hole. An alternative to this is to fashion a rectangular guide for the vertical yardstick and attach the guide to the horizontal stick.

A more elegant rate tester can be fashioned from a large cork, some wire, and a short ruler (or even a regular 1-foot ruler). The ruler is fastened in a vertical position to the horizontal stick. Roughly half the ruler is below the stick and the balance is above.

One end of the wire is driven through the cork and beyond for several inches. The upper end of the wire is fashioned into a horizontal rectangle that will pass loosely around the top of the ruler. The wire extending below the cork is curled into a ball. The cork and wire are now lowered onto the water in the hole. The rectangle of wire passes around the ruler. The curled wire should be heavy enough to keep the device vertical. If it is not, add some more wire for weight. Now you can read the position of the rectangular wire loop against the markings on the ruler.

Timing. Using any of the devices suggested previously, adjust the starting level of the water in the hole to bring whatever indicating method you use to an inch mark (just for easy reading). With a watch or stop watch, if the time is short, measure the time required for the water to drop exactly one inch. This is the figure you are seeking: time per inch of drop.

One minute might be required or an hour might be required. This is the figure you need and it is crucial.

After the first test in hole one, wait several minutes (or go to the second hole) and then repeat the test. Do this three times or more, if the results vary widely, until the figures are relatively alike.

FINDING THE REQUIRED DRAIN-FIELD SIZE

When you have repeated the tests a sufficient number of times to secure relatively identical readings, you can be reasonably certain you have the figure (time) you need. Identical readings from repeated tests of the same test hole indicate the soil has become stabilized; it has absorbed a quantity of water, duplicating, more or less, actual operating conditions. You can now use this time figure to find the drain field pipe length you require. Just refer to Table 20-1.

Computing Drain Field Size. First of all, let me emphasize that the following example refers to minimum figures. Increasing the length and thus the capacity of the system provides a measure of safety against errors and a margin for future increased usage. No harm is done excepting to one's pocketbook.

Knowing the percolation rate provides you with the drain pipe trench-wall area factor. Having determined the best depth for the drain pipe in your area, you now know how far down the center line of

Table 20-1. Time to Drain Versus Required Drain-Pipe Wall Area.

Rate	Recommended minimum drain pipe wall area
1 minute or less	1.0 sq ft/gal/day*
2	1.5
3	2.0
4	2.5
5	3.0
10	4.0
20	4.6
30	5.2
40	7.0
60	9.-
Unsuited	----

*This figure is for the "active" sides of the trench only. It does not include the bottom of the trench nor the area above the centerline of the drain/pipe, and it includes only one side of the two walls of a trench. If you have a trench 10 feet long and the pipe was positioned with its centerline just 1 foot above the bottom, that trench would have 20 square feet of active wall area or surface. Note: You might find this table at variance with other published tables on the same subject. These figures are a compromise.

the pipe will be from grade. Now you have to decide how much farther you want to dig. Remember the effective wall area of the trench extends only from the centerline of the drain pipe. See Fig. 21-4.

Assume that you decide to make your trench 28 inches deep and position the pipe's top 12 inches from grade. This leaves a sidewall effective distance of 14 inches. To find the active area of this

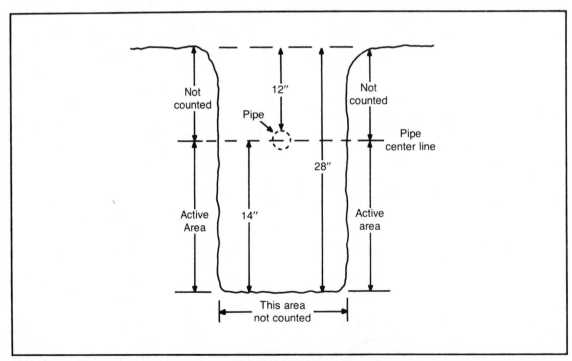

Fig. 20-4. How the active (useful) area of the walls of a drain trench are measured. The active area in this example is 2 times 14 inches for every foot of trench length. This gives an answer of 2 square feet.

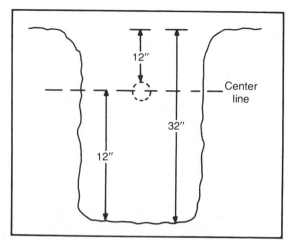

Fig. 20-5. In this example, the active area is only 12 inches high. This makes for a big difference in the total trench length required.

trench per lineal foot of length, multiply:

14 × 12 × 2 = 336 square inches

336 ÷ 144 = 2.4 square feet/lineal foot

Assume that the daily effluent outflow averages 650 gal/day. To find how many square feet of wall area is required, divide.

650 ÷ 5.2 = 125 square feet (area for 30 min rate)

To find how many lineal feet of trench is needed divide:

125 ÷ 2.4 = 52 lineal feet of trench needed.

Let's take another example (Fig. 20-5). Assume the perc test shows 10 minutes. That time indicates 4 square feet of drain wall surface is required. Suppose the effluent figure, 650, remains the same. Now instead of making the trench 28 inches deep, you make it 32 inches deep, still keeping the top of the drain pipe 12 inches below grade. That leaves 18 inches of active wall surface.

18 × 12 × 2 = 432 square inches

432 ÷ 144 = 3 square feet per lineal foot of trench

650 gal/day ÷ 4.0 (the 10 minute rate) = 163.

To find how many lineal feet of this trench you need, divide:

163 ÷ 4.4 = 37 feet.

As you can see, the perc rate of the soil plus the depth of the trench below the drain pipe center line have together an enormous affect on the total drain pipe required. In the first example, you needed 52 lineal feet of pipe. In the second example, only 37 feet of trench is required. The difference is due in part to the faster absorption rate (10 minutes in place of 30 and the deeper trench 32 inches in place 28 inches) helped shorten the drain pipe.

Explanation of Active Area. The area comprising the bottom of the drain pipe trench is not included in absorption computations because the bottom of the trench is the first area to be "plugged" up by bacterial growth and debris. While this area will participate in effluent absorption for the first few months of system operation, the area will gradually lose its ability to absorb effluent and become useless.

The area above the center line of the drain pipe is not considered in the absorption calculations because the effluent rarely, if ever, rises above this height in the trench.

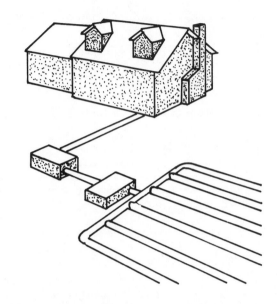

Chapter 21

Constructing the Drain Field

PRECEDING CHAPTERS DISCUSS THE THEORY of septic tank operation and the conversion of the effluent flowing from the tank into harmless water by aerobic bacteria. Methods and calculations by which you can determine the required length of your drain field pipe are also covered. At this point, you know the total length of the drain field pipe you require and you also know the depth to which the pipe will be positioned in the trench and the depth of the trench.

The pipe leading from the septic tank to the drain field pipe, which is perforated, must be solid. You do not want effluent seeping from the "feed" pipe to the surrounding earth.

Use only plastic pipe or clay drain tile. Do not use Orangeburg Pipe (a mineralized paper pipe). It will quickly deteriorate.

Insulate the feed pipe where it will pass beneath a driveway. Cover it with batts of glass wool. There is less soil above the feed pipe at this point and driveways hold the cold.

If possible, run the drain pipe in a straight line. The fewer turns and the fewer branches the less

likely solids in the effluent will get hung up., It also will be easier to hold the pipe to the proper pitch.

Separate parallel drain pipes by a distance 1½ times the effective (active) depth of the pipe. For example, if the distance from the proposed center line of your pipe to the bottom of the trench will be 2 feet, the nearest parallel pipe should be 3 feet. That is the trench side wall and not the side of the pipe within the trench.

Do not place the drain field within a natural depression where rain water collects and remains a long time after the rain has stopped falling.

EXCAVATING THE TRENCH

You can excavate with the time honored method of using strong back and trusty shovel. You can hire a machine with a giant wheel and buckets on its edge called a *trencher*. When the wheel is lowered against the earth and revolved, a trench is excavated in short order. This machine is preferred to the *back-hoe* because the trencher is faster, cuts a cleaner-sided hole and usually cuts a narrower trench. The backhoe removes earth in single gobs, generally a trench 24 inches or wider, and that might or might not be what you want.

Trench Depth. Keep constant track of trench depth with a yardstick or folding rule. If your shovel is hand powered, you are wasting muscle power when you dig too deeply. When a machine is doing the work, too shallow a depth means you will have to deepen the trench by hand. Too deep is not good, either, but it can be corrected by simply adding more gravel.

Trench Width. Trench width is generally 18 inches or more. Less than this limits the quantity of

gravel that can be placed alongside. A greater width is usually not particularly beneficial. By rule of thumb, trench width should, at a minimum, be one-third that of the overall depth of the trench.

Sides and Bottom. Trench sides should be kept vertical if possible, but this is only a convenience. Pitched sides just mean more soil to remove. Each trench section should be kept to a straight line so as to keep the drain pipe more or less centered in its trench.

The trench bottom should be reasonably flat. So long as the bottom remains at or below the required depth, there is no need to seek perfection.

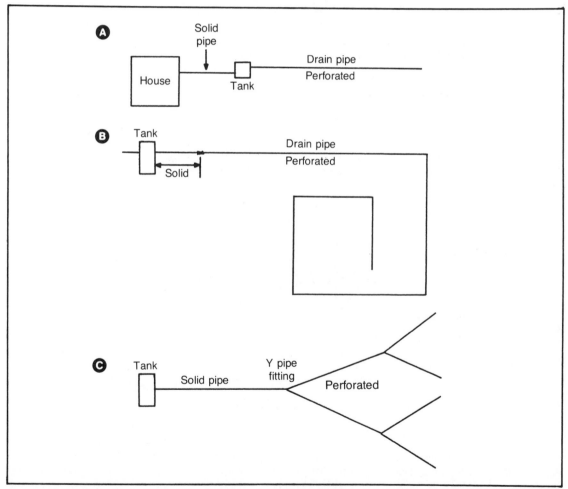

Fig. 21-1. The pipe carrying the effluent from the tank to the drain field should be solid until the pipe becomes the drain field. A, B and C show three ways of running drain field pipe. The straight line is the best because the fewer joints and angles the less chance of hangup.

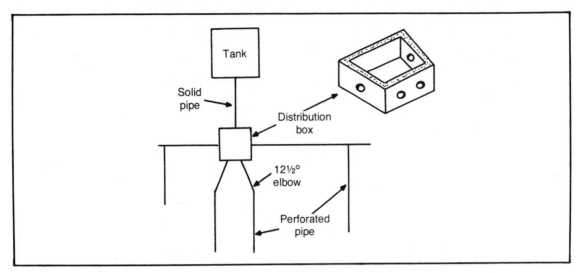

Fig. 21-2. Another way of running a drain field. Here the division of the effluent is made in a distribution box. This box should be inspected once a year.

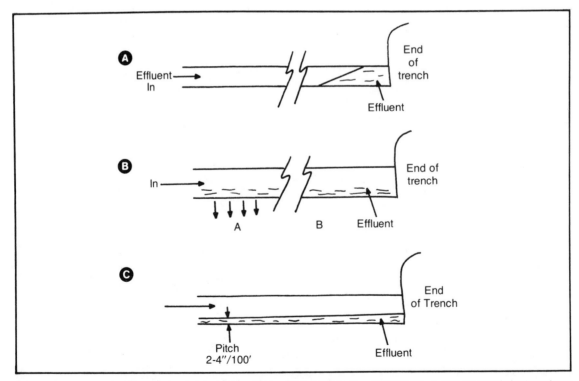

Fig. 21-3. How the pitch of the perforated drain pipe affects effluent distribution and field life. A: The pipe is pitched much too steeply. All of the effluent runs to one end. As a result, this portion of the pipe is overloaded and soon becomes plugged. B: Here the pipe is perfectly level. The effluent divides evenly at first, but because the effluent spends more time at the start of the pipe this portion is overloaded and plugs up first. C: Here the pitch is gentle. A little more effluent flows to the end of the pipe than to its start, but that results in even use by the pipe and surrounding soil.

222

Layout. You can run the drain pipe in a straight line, you can make the pipe line fold back upon itself several times, and you can divide the pipe into any number of branches. Just bear in mind that the more turns and branches you include the more difficult it will be to excavate by machine and the more difficult it will be to hold the pipe to the necessary pitch. See Fig. 21-1.

Branches. You can divide the drain pipe into two or more branches with standard sanitary Ts and Ys. When several branches are to emanate from a single pipe, it is best to use a distribution box rather than a number of Y fittings. See Fig. 21-2.

The distribution box is a concrete box more than large enough to accept all the pipe ends entering and leaving. The boxes are available already fabricated from commercial sources, or you can construct a form and pour your own. In the latter case, make the walls 3 inches or more and round the inside corners so that there are no angles (angles make annual cleaning difficult). The box should be provided with a heavy concrete cover and positioned well within the earth to protect the effluent against frost.

SECURING THE REQUIRED PITCH

The solid pipe leading to the drain field perforated pipe can be run at any angle from ⅛ inch per foot on up. The drain pipe pitch requires careful consideration and execution. See Fig. 21-3.

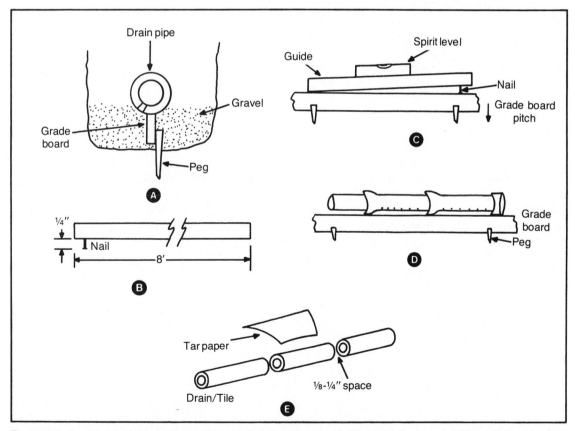

Fig. 21-4. A: An end view of a drain pipe resting on a grade board surrounded by supporting gravel. B: The guideboard and the nail that provides the desired pitch of 2 inches in 100 feet. C: When the guide is on the drain pipe, nailhead down, and the spirit level indicates the board is level, you have the desired pitch. D: With short sections of pipe, you have to take care to make certain the bells of succeeding sections of pipe are all on the gradeboard. E: Here is how the short sections of drain tile are handled. The gradeboard and supporting gravel are used (though not shown in this drawing).

The theory behind the pitch is that you want the effluent to run from one end of the pipe to the other without filling the end of the pipe, and without too high a percentage of the liquid collecting in the second half of the pipe. In other words, you want the effluent to linger an equal time over the entire length of the drain pipe. Generally, the pitch used ranges from 2 to 4 inches per 100 feet of length.

In highly permeable soil, you should opt for the greater pitch because some of the effluent will be absorbed by the first section of drain pipe. To some extent, this will overload the first section and leave the balance relatively dry. In low-permeable soil, the lower pitch would work out better.

It is easier to control pitch when laying plastic pipe because the individual sections are long. It is more difficult with drain tile because they are so short. In either case, the only practical way to hold to pitch is with the aid of a grade board (actually several of them). Use 2-×-4 fir (the least expensive wood) in as long and straight-edged lengths as you can find.

Nail several short pegs to each length of fir. Drive the pegs into the earth in the center of the trench. Doing so will position the grade board on edge at the bottom center of the trench.

Now you need to pitch the gradeboard at the desired angle. This can be accomplished with a homemade guide. Drive a small nail into the edge of an 8-foot, straight-edged board, near the end of the board. Tap the nail until its head is just ¼ of an inch from the board edge. Place this board on edge and on the nailhead on a flat surface. The guide will have a pitch of close to 2 inches in 100 feet.

To use this guide, it is positioned nailhead down on the grade board). A spirit level is placed atop the guide. The grade board is moved up or

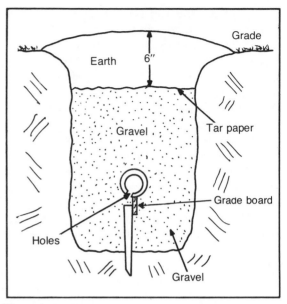

Fig. 21-5. The gravel surrounds the drain pipe and the earth is mounded so that it will eventually compact level with the balance of the soil.

down until the bubble in the spirit level indicates level. At that point the grade board is at the 2-inch-per-100-foot pitch.

With the gradeboard at the desired pitch, gravel is placed beneath it and to the sides to hold the board on pitch. The drain pipe or tile is now laid atop the gradeboard. The balance of the gravel is positioned and the grade board remains in place. See Fig. 21-4.

Completion. The trench is filled with gravel, 6 or so inches short of the top of the trench. A layer of tar paper is then placed over the gravel, followed by earth. See Fig. 21-5. The earth is mounded for a height of 4 to 6 inches. In time, the earth will compact and the mound will disappear.

Glossary

aeration, zone—Formation extending from the water table to the surface of the earth in which no free water is found.

aerobic bacteria—Bacteria that do not thrive without oxygen.

anaerobic bacteria—Bacteria that thrive in the absence of oxygen.

annular velocity—Speed of air or water through an annulus.

annulus—Circular space between a central pipe and a surrounding casing.

bailer—A special, long pipe with a value of its bottom used to remove sand and cuttings from a bore hole.

Bentonite—A kind of clay used in drilling.

black water—Water from a toilet.

bottom-hole drill—A drill that is connected to a motor that fits into the bore hole.

cable-drill tool—A tool(s) that are lowered into the bore hole on the end of a cable.

casing shoe—An edged collar fastened to the lower end of a well casing that is to be driven.

cat head—A slowly revolving smooth-surfaced drum.

centrifugal pump—A pump that drives water by means of the rotary action of its blades.

cuttings—Chips of the formation produced by the drill bit.

dart valve—A fast-acting valve sometimes used on the bottom of a bailer.

developing—Removing the fine sand and silt from the area immediately surrounding the well screen.

down-hole hammer—A rotating, reciprocating bit driven by an air hammer (both are lowered to the bottom of the bore hole).

drawdown—The lowering of the static water level by pumping the well.

drill stem—One of the tools making up a drill string.

drive pipe—Pipe attached to a well point.

effluent—The liquid that emerges from a septic tank.

formation—A general, geological description of the subsurface earth and stone.

grey water—Sewage from sinks and tubs that does not include water from toilets.

grout—Liquid cement, sometimes liquid cement and sand.

head drive—When the motor driving the drill is mounted on the mast.

hydraulic rig—When the drive motor is actuated by hydraulic power.

impeller pump—A type of propeller pump.

jars—Two loose fitting parts making up a drill string producing an upward hammering action when the cable is raised.

jet lance—A pipe through which water is forced.

jetting—Using a powerful stream of water to cut a bore hole into the earth.

kelly—A shaped- or splined-steel bar that passes through a rotary table and is turned by the table.

lead packer—A shaped ring of lead that is fastened to the top end of a well screen. When expanded within a casing, the screen is sealed to the casing.

mast—A vertical tower that carries the sheave from which the cable tools hang. Can also carry head-drive motors and auxiliary sheaves.

meander tank—A septic tank designed to provide the longest practical path for the effluent.

mud scow—A system of excavating a bore hole in a soft formation with down pressure on the casing and a jet stream of water.

packing, artificial—The gravel placed around a well screen.

packing, natural—The gravel around a well screen after the well has been developed.

pathogen—Harmful organisms invisible to the naked eye.

pitless adapter—A device for connecting a water line to a well and leaving access to the well above ground.

precussion drilling—The system of raising and lowering a drill string so that the attached drill bit pounds a hole in the formation.

pull down—Pressure applied to a casing to force it into the formation (as distinct from pounding or driving).

recharge—Rain or other water that enters an aquifer.

rig—Equipment used to drill or bore wells.

riser—Special pipe connected to a well point.

rotary table—A flat, circular disk with a shaped hole to accept the kelly. When the disk is rotated, the kelly turns with it.

sand line—A light cable used to raise the bailer.

sand pump—A pump mounted inside a bailer to lift sand from the bottom of the bore hole.

saturation zone—The area in an aquifer that contains free water.

seep—A natural spring just a few inches below the surface of the earth that keeps an area constantly wet.

siphon tank—A septic tank containing a siphon for fast movement of the effluent from one tank section to another or to the drain pipe.

spudding—The up and down action of a precussion drill; the mechanism providing this action.

static water level—The water table when the well is not being pumped.

suction head—Total negative pressure required to lift water from the source to the pump center, in feet.

termie—A pipe passed down an annulus through which gravel or grout can be forced to the bottom of the bore hole.

vacuum pump—A pump that raises water by suction (vacuum).

wash down—Using a stream of water under pressure to provide a hole in the earth beneath a casing.

water table—Upper surface of ground water within an aquifer before any pumping is done.

well casing—A steel pipe that might contain a second pipe within itself that carries the well water. The casing itself might carry the well water.

well pipe—The pipe that carries the well water.

Index

Edited by Steven Bolt